Gas Turbine Parameter Corrections

Allan J. Volponi

Gas Turbine Parameter Corrections

Allan J. Volponi
Pratt & Whitney
West Simsbury, CT, USA

ISBN 978-3-030-41078-0 ISBN 978-3-030-41076-6 (eBook)
https://doi.org/10.1007/978-3-030-41076-6

This Springer imprint is published by the registered company Springer Nature Switzerland AG
The registered company address is: Gewerbestrasse 11, 6330 Cham, Switzerland

Dedicated to the late Louis A. Urban, my long-time friend, colleague, and mentor, who introduced me to my first corrected parameter.

Preface

Parameter corrections are a familiar and ubiquitous practice in describing interrelationships within the gas turbine engine's flow path. A frequent usage is to remove dependency on engine inlet ambient conditions (primarily temperature and pressure) to allow succinct relationships to be developed and employed in a variety of applications. Nevertheless, the genesis of these corrections, and in particular, how they can be derived from simple principles, is not general knowledge among many of those who use them on a day-to-day basis. This book is intended to help fill that gap.

With the exception of a single chapter on humidity corrections, this text concerns itself with what is commonly referred to as *Standard Day* corrections of gas turbine gas path parameters. This type of correction has been in use for over five decades, and as such is common currency in the engineering community engaged in a variety of activities, most notably, engine performance specifications and calculations, control law development, modeling and engine diagnostics, prognostics and health management, to name a few. Due to its longevity and common knowledge across a diverse gas turbine engineering community, it is *not* the intent of this work to explore the many uses of these familiar corrections, but rather to explore their history, provide an exhaustive catalog of common (and some not so common) corrections, and most importantly to provide a mathematical framework for the derivation of these important normalization factors. Little reference to application will be made. A large part of this work will be devoted to the derivation of these corrections with the intent that this will serve as a handy desk reference to the engineering practitioners utilizing these corrections in their everyday activities.

West Simsbury, CT, USA Allan J. Volponi

Abstract

This book concerns itself with what is commonly referred to as *Standard Day* corrections of gas turbine gas path parameters. Due to its longevity and common knowledge across a diverse gas turbine engineering community, it is *not* the intent of this work to explore the uses of these familiar corrections, but rather to explore their history, provide an exhaustive catalog of common corrections, and most importantly to provide a mathematical framework for the derivation of these important normalization factors. Little reference to application will be made. We include here, a short Preface and Introduction describing the layout of the book.

Keywords Introduction · Structure · Chapters · Description

Introduction

It is well known that, at a given operating point, gas turbine gas path parameters such as inter-stage pressures, temperatures, and flows will vary with ambient conditions. As such, it is common practice to normalize these parameters for ambient temperature, pressure (and to a lesser degree, humidity) before performance calculations are preformed, especially in cases where performance is to be trended over a long period of time when it is likely that these conditions will vary. The normalization methodology is commonly referred to as *Standard Day* corrections. Like many engineering practices, it is not a (mathematically) precise method, but rather an *approximation* that is sufficient for the purpose under which it is used, which is primarily to remove the variation induced solely by the ambient environment in which the gas turbine operates. These corrections are pervasive, in the sense that they have been applied in virtually every gas turbine engineering activity ranging from design and development of components and sub-systems, control strategies and schedules, performance calculations as well as engine monitoring, trending, and health management activities. In short, their existence and routine application are well known and well utilized [1–4]. As such, we will *not* venture into the broad field of applications exploiting these corrections. The purpose of this book is to establish a (quasi-mathematical) framework from which to establish these corrections and explore certain interrelationships. The use of the term, quasi-mathematical, is intentional in that the formal mathematical manipulations that will be used in our discussion arise from certain stated relationships, some thermodynamic, some not, that will be assumed to be the axioms from which our results will be derived. We will come back to this point several times throughout this presentation.

With the increased use of computer simulations, the need for such correction calculations has decreased in recent time over what was needed decades ago when analytical methods prevailed and formed the only path to explore performance assessment and manipulate data. Nevertheless, the use of parameter corrections is still an important consideration especially in real-time engine modeling, control scheduling and strategy, and by researchers and practitioners who do not possess the specialized, and often proprietary, engine simulations that are the property of engine designers and manufacturers. For these reasons, knowledge of parameter corrections is still an important, if not vital, factor in supporting engineering research and practice. Reference to corrected parameters and their usage can be found from many sources, books, papers and the like, albeit incompletely. *This book seeks to put in one place a compendium of all known corrections (and some perhaps not so well known) as well as to provide a mathematical foundation for their derivation.*

The presentation begins in Chap. 1 with an historical perspective containing a short summary of dimensional analysis and the use of *Buckingham's π Theorem* as a motivation for how corrected parameters originally evolved. This methodology is a bit tedious and not well suited for deriving parameters of interest, especially those involving time derivatives. It is, however, the genesis of what we know today as *Standard Day* corrections and is included for completeness. Chapter 2 introduces a more general mathematical foundation to overcome some of the shortcomings of pure dimensional analysis and stipulates a handful of intuitive axioms (assumptions) to allow derivations to ensue armed with only an elementary knowledge of thermodynamics and Newton's calculus. Using these principles, Chap. 3 derives the corrections for most common gas path parameters with Chap. 4 extending to the time derivatives of those parameters. Chapter 5 explores some additional corrections that follow easily from the derivations of Chaps. 3 and 4. In Chap. 6, the *assumption of constant specific heats is relaxed* to refine the corrections to several of the most common gas path parameters. The tone of the text changes in Chap. 7 with the introduction of empirical methods for further refinement, and we conclude our derivations in Chap. 8 with a discussion of the effects of humidity on performance parameters. The Appendix summarizes all of the corrections derived in this book.

Lastly, a few comments on symbolic notation. The symbol ($\Rightarrow$) appears throughout the text in many of the derivations. This symbol means *which implies* or *implying*. Since it is understood that the corrections are by their nature an approximation, we will use the symbol approximately equal ($\approx$) interchangeably with mathematically equal ($=$) throughout our presentation as a matter of convenience.

Contents

Nomenclature

Symbol

A	Area
α_X	Theta (θ) exponent for gas path parameter X
cp	Specific heat at constant pressure
$cp_{a,\ d}$	Specific heat of dry air at constant pressure (same as cp)
$cp_{a,\ w}$	Specific heat of wet air at constant pressure
cp_w	Specific heat of water vapor at constant pressure
cv	Specific heat at constant volume
δ	Pressure ratio P/P_{ISA}
EGT	Exhaust gas temperature
EPR	Engine pressure ratio
η_b	Burner efficiency
FAR	Fuel–air ratio
F_n	Net thrust
g	Gravitational acceleration constant
h	Enthalpy
HP	Horse power
γ	Ratio of specific heats
I	Moment of inertia
ISA	International Standard Atmosphere
J	Mechanical equivalent of heat
LHV	Fuel lower heating value
M_n	Mach number
μ_X	Humidity correction factor for gas path parameter X
N	Rotor speed
$\dot{N}$	Rotor acceleration
P	Pressure (total)

P_s	Pressure (static)
P_{ISA}	Sea level ISA pressure
PLA	Power lever angle
Q	Torque
R	Gas constant
$R_{a,\ d}$	Gas constant of dry air (same as R)
$R_{a,\ w}$	Gas constant of wet air
R_w	Gas constant of water vapor
r	Radius
ρ	Density
SH	Specific humidity
$TSFC$	Thrust specific fuel consumption
T	Gas temperature (total)
T_s	Gas temperature (static)
T_m	Metal temperature
T_{ISA}	Sea level ISA temperature
τ	Time constant
Θ	Temperature ratio T/T_{ISA}
S	Entropy
v	Velocity
V	Volume
w_a	Air mass flow
w_f	Fuel mass flow
war	Water to air ratio

Subscript

a	Air
c	Corrected (parameter)
f	Fuel
d	Dry
w	Wet
s	Static
m	Metal property
HD	Hot day

Chapter 1
Historical Perspective

Abstract This chapter provides some historical perspective contributing to the use of gas path parameter corrections. It largely involves the use of dimensional analysis and the so-called π theorem due to Edgar Buckingham, an American physicist working at the U.S. National Bureau of Standards in the early twentieth century. This chapter briefly describes the π theorem principle and illustrates the concept with several examples. The reduction in complexity obtained by applying this principle to yield equivalent functional relationships involving new variables (that are rational functions of the fundamental variables) was the driving force that ultimately formed the parameter correction process in common use today.

Keywords Buckingham · π-theorem · Dimensional analysis · Historical · Procedure · Complexity reduction · Dimensionless products · Examples

There is an important historical perspective that originally gave rise to the type of corrections we will be considering that bears consideration. It is sometimes referred to as *dimensional analysis*[1] and has its roots in the early part of the twentieth century with the work of Buckingham [5]. We will begin with a brief description of his work as a precursor to our formal development.[2] Before the advent of digital computers, engineering calculations were tediously performed by hand with the use of simple devices such as slide rules and the like. Under these conditions, it is advantageous (for time and effort) to reduce the complexity of any calculation that needed to be performed. Reducing the number of independent variables in physical interrelationships is one way to reduce complexity. This is at the core of Buckingham's work and resulted in a relationship that has come to be known as Buckingham's π theorem. The following description is taken largely from [6].

---

[1]See [15].

[2]The interested reader may wish to consult [13, 14] for the use of non-dimensional parameters in gas turbine propulsion and thermodynamics.

A. J. Volponi, *Gas Turbine Parameter Corrections*,
https://doi.org/10.1007/978-3-030-41076-6_1

Let us assume that we have a physical relationship to be represented by n physical quantities Q_1, Q_2, …, Q_n. Without any loss of generality, we may write this relationship as

$$f_1(Q_1, Q_2, \ldots, Q_n) = 0 \tag{1.1}$$

Since physical quantities are measured in physical units (feet, seconds, lbs, degrees, etc.) or some algebraic combination of units, it can be shown that the above general relationship must have the form

$$\sum MQ_1^{b_1} Q_2^{b_2} \ldots Q_n^{b_n} = 0 \tag{1.2}$$

where M is a dimensionless number and each of the products being summed must have the same units (otherwise you could not add them together). If we divide both sides by any one term in the sum, the expression takes the form

$$\sum NQ_1^{a_1} Q_2^{a_2} \ldots Q_n^{a_n} + 1 = 0 \tag{1.3}$$

where each product in the sum is a dimensionless quantity. If we represent each of these terms by π_i, then we may rewrite this equation as

$$f_2(\pi_1, \pi_2, \ldots, \pi_n) = 0 \tag{1.4}$$

Loosely stated, the Buckingham π Theorem declares that the number of dimensionless products π_i necessary for Eq. (1.4) to represent Eq. (1.1) is $n - k$, where k is the number of fundamental dimensions used to define the variables.

To illustrate this principle, let us consider the problem of determining the distance travelled by a free falling object (in a vacuum) and derive a relationship for this distance (without the aid of Mr. Newton's calculus) using dimensional analysis and Buckingham's π Theorem. We will assume that the variables under consideration for this problem are given in the Table below, where the units are length (L), mass (M), and time (T).

Parameter	Description	Units
S	Distance traveled	L
W	Weight of object	$\mathrm{M\,L\,T^{-2}}$
g	Gravitational constant	$\mathrm{L\,T^{-2}}$
t	Time	T

We will assume the following general relationship holds:

$$f(S, W, g, t) = 0 \tag{1.5}$$

We have four parameters and three unique fundamental dimensions (L, M, T). By Buckingham's theorem, we have $4 - 3 = 1$ π terms as follows:

$$\begin{aligned}
\pi_1 &= KS^a W^b t^c g^d \Rightarrow \\
M^0 L^0 T^0 &= (L^a)(M^b L^b T^{-2b})(T^c)(L^d T^{-2d}) \Rightarrow \\
&= L^{a+b+d} M^b T^{c-2(b+d)} \Rightarrow \\
b &= 0 \\
a + b + d &= 0 \\
c - 2b - 2d &= 0
\end{aligned} \tag{1.6}$$

From the last three equations, we find that

$$\begin{aligned}
a &= -d \\
c + 2a &= 0 \Rightarrow \\
a &= -c/2
\end{aligned} \tag{1.7}$$

Since g is a constant, we can take $d = 1$,[3] and then we may conclude that

$$\begin{aligned}
a &= -1 \\
b &= 0 \\
c &= -2a = 2 \Rightarrow \\
\pi_1 &= KS^{-1} W^0 t^2 g = K\frac{gt^2}{S}
\end{aligned} \tag{1.8}$$

Now since π_1 is dimensionless, it must be a constant (as is K), therefore

$$S = \left(\frac{K}{\pi_1}\right) gt^2 \tag{1.9}$$

Experimentation would suggest that the constant $K/\pi_1 = 1/2$, providing the well-known relationship from Physics 101 without the aid of Calculus. An interesting side note, is that the dimensional analysis suggests that the mass (M) of the object has no bearing on the interrelationship of the freely falling body, a fact that was demonstrated experimentally and dramatically by Galileo in Pisa, circa 1590, and later by the American astronaut Col. David R Scott on the lunar surface in 1971!

So how does this help us with correcting gas turbine parameters? We will illustrate this with another simple example from [2]. Suppose that we have a single spool turbojet engine having the following gas path parameters:

[3] This choice is arbitrary since any value will still provide the same result. The reader can easily verify that choosing $d = 2$, for instance, will result in the same relationship noted in Eq. (1.9).

Parameter	Description	Units
N	Spool speed (rpm)	1/T
T_1	Ambient temperature	deg
P_1	Ambient pressure	M/(LT^2)
v	Aircraft velocity	L/T
D	Compressor diameter	L
R	Gas constant	L^2/(T^{2}deg)
F_n	Net Thrust	M L/T^2

where T = Time, deg = temperature (degrees), M = Mass, and L = Length. The total number of fundamental dimensions in this case is 4. We assume that we have the following relationship:

$$Fn = f(N, T_2, P_2, v, D, R) \tag{1.10}$$

This implies an implicit form

$$f(Fn, N, T_2, P_2, v, D, R) = 0 \tag{1.11}$$

involving seven quantities. By virtue of the π-theorem, we may express the relationship in terms of 7 – 4 = 3 dimensionless quantities as follows:

$$\begin{aligned} &\phi(\pi_1, \pi_2, \pi_3) = 0 \quad \text{where} \\ &\pi_1 = NP_2^a T_2^b D^c R^d, \\ &\pi_2 = vP_2^e T_2^f D^g R^h, \\ &\pi_3 = F_n P_2^k T_2^m D^n R^r \end{aligned} \tag{1.12}$$

The π_i terms are dimensionless and hence their units must be equal to $L^0M^0deg^0T^0$. For example

$$\begin{aligned} \pi_1 &= \left(\frac{1}{T}\right)\left(\frac{M}{LT^2}\right)^a \deg^b L^c \left(\frac{L^2}{T^2\deg}\right)^d = L^0M^0\deg^0T^0 \\ &= L^{c+2d-a} M^a \deg^{b-d} T^{-1-2a-2d} \\ &\text{which implies} \\ 0 &= a \\ 0 &= c + 2d - a \\ 0 &= b - d \\ 0 &= 1 + 2a + 2d \text{ which implies} \quad a = 0 \\ &\qquad\qquad b = d = -1/2 \\ &\qquad\qquad c = -1 \end{aligned} \tag{1.13}$$

This yields a parameter of the form

$$N/\left(L\sqrt{RT_2}\right) \tag{1.14}$$

Repeating this process for the other dimensionless quantities, yields the following relationship

$$\frac{F}{P_2A} = f_1\left(\frac{N}{D\sqrt{RT_2}}, \frac{v}{\sqrt{RT_2}}\right) \tag{1.15}$$

If we assume a constant geometry (A = const, D = const) we obtain

$$\frac{F}{P_1} = f_2\left(\frac{N}{\sqrt{T_2}}, Mn\right) \Rightarrow \frac{F}{\delta} = f_3\left(\frac{N}{\sqrt{\theta}}, Mn\right) \tag{1.16}$$

This would imply that the appropriate (and well known) corrections for net thrust and spool speed are F/δ and $N/\sqrt{\theta}$ respectively.

Now strictly speaking, we have only demonstrated these relationships for a single spool turbojet engine with constant geometry and thus would have to repeat the type of dimensional analysis for the specific engine under consideration (if different). What is needed is a more general approach to demonstrate that these corrections can be applied in general to *any* given gas turbine configuration. With this in mind, we will offer a more general *definition* of what a *corrected* parameter should represent and derive a general form for the correction.

Chapter 2
Mathematical Framework

Abstract This chapter explores the construction of a mathematical framework to enable the direct derivation of Standard Day corrections for an arbitrary gas path parameter. The fundamental approach is in contrast with the historical methodology of dimensional analysis (briefly) mentioned in Chap. 1. This is accomplished by establishing a formal definition of a *corrected* parameter along with a short list of properties that corrected parameters are assumed to possess. These collectively form a set of axioms for our mathematical system from which formal derivations can ensue, using only fundamental thermodynamic principles and elementary calculus. This framework will provide the basis for the discussions contained in all subsequent chapters of this book except for Chap. 8, which deals exclusively with the effects of humidity on performance.

Keywords Corrected parameter · Properties · Definition · Axioms · Standard Day Correction · Mathematical framework · Notation · Delta · Theta · Engine stations · Hot Day · Reference conditions · Assumptions · Specific heat · Local Mach No. · Corrections of parameter square root · Corrections of parameter raised to arbitrary power

In this chapter we will develop a mathematical framework to allow derivation of any corrected gas path parameter independent of gas turbine engine type or configuration.[1] To accomplish this we will introduce some terminology and propose a set of fundamental axioms that will be utilized in the derivations appearing throughout the remaining chapters of this book.

For any gas path parameter X, the equivalent *corrected* parameter will be denoted by X_c throughout this discussion. In general, a change in the engine inlet conditions T_2 (inlet total temperature) and P_2 (inlet total pressure) will be accompanied by an attendant change in any downstream gas path parameter X. A *corrected* parameter X_c would be *constant regardless of the change in inlet condition* and represents the

[1]We will assume fixed geometry.

A. J. Volponi, *Gas Turbine Parameter Corrections*,
https://doi.org/10.1007/978-3-030-41076-6_2

value the parameter X would have at some *fixed reference inlet condition*. This reference condition can be whatever you like, however, it is common practice to select the International Standard Atmosphere (ISA) *standard day* conditions ($T_2 = 288.15$ deg K, $P_2 = 101325.353$ Pa, or alternatively, $T_2 = 459.67$ deg R, $P_2 = 14.696$ psia) for this purpose. Thus, we can assume without loss of generality that a *corrected* parameter quantity X_c is one that satisfies the following *implicit* relationship.

$$\text{Axiom 1}: \quad X = f(X_c, T_2, P_2) \tag{2.1}$$

The interpretation of Eq. (2.1) requires some explanation. Its intent is that a gas path parameter X will vary from its *corrected* form X_c **only** on account of changes in ambient condition as given by the inlet temperature and pressure. Implicit in this statement is that something is being held constant. For example, if power condition were to change (e.g., through a change in Power Lever Angle, PLA), then that would certainly affect X as well. Indeed, so would X_c. So in some sense, we are assuming from this implicit definition that the *state* of the engine has been held fixed. But what does this mean? Whatever it is, it needs to be independent of the engine control system, otherwise corrected parameters would *not* be independent of engine model, configuration and method of control. So we can rule out things like Engine Pressure Ratio (EPR) or Thrust, etc. What is certain is that we need for the thermodynamic and aerodynamic *condition* of the engine to be fixed. Since the gas turbine is an air breathing machine and air is the elemental fluid whose properties are being measured by the gas path parameters that we wish to normalize (i.e., *correct*), *we will assume that local Mach numbers along the gas path have remained fixed.* Consider, for example, Airflow (w_a), which can be seen to be solely a function of geometry (fixed), Mach number, and (Total) Temperature and Pressure, i.e.,

$$\begin{aligned}\frac{w_a\sqrt{T_2}}{AP_2} &= \sqrt{\frac{\gamma}{R}}\frac{M_n}{\left(1+\frac{\gamma-1}{2}M_n^2\right)^{\frac{\gamma+1}{2(\gamma-1)}}} = f(M_n) \Rightarrow w_a = f(M_n, T_2, P_2) \\ &= f(T_2, P_2)\end{aligned} \tag{2.2}$$

With M_n being held fixed, airflow will be governed only by total temperature (T_2) and pressure (P_2) as needed. Similar arguments can be given for other gas path parameters. Thus it would appear that holding local Mach numbers constant to be a reasonable condition to preserve a fixed engine *state*. We will apply this assumption liberally throughout this discussion and make other simplifying approximations (e.g., averaging) where needed. Since the corrections we are deriving should only be considered *approximations* themselves and not precise physical quantities, we can afford ourselves these conveniences.

Equation (2.1) will be our working definition for what a corrected quantity represents. Through formal differentiation, it follows that

$$dX = \left(\frac{\partial X}{\partial T_2}\right)_{\substack{P_2=const\\X_c=const}} dT_2 + \left(\frac{\partial X}{\partial P_2}\right)_{\substack{T_2=const\\X_c=const}} dP_2 + \left(\frac{\partial X}{\partial X_c}\right)_{\substack{P_2=const\\T_2=const}} dX_c \text{ implying}$$

$$\frac{dX}{X} = \left(\frac{\partial X/X}{\partial T_2/T_2}\right)_{\substack{P_2=const\\X_c=const}} \frac{dT_2}{T_2} + \left(\frac{\partial X/X}{\partial P_2/P_2}\right)_{\substack{T_2=const\\X_c=const}} \frac{dP_2}{P_2} + \left(\frac{\partial X/X}{\partial X_c/X_c}\right)_{\substack{P_2=const\\T_2=const}} \frac{dX_c}{X_c}$$

If we make the (brash) *assumption* that the *partials are approximately constant* (the third is clearly unity by definition) we have

$$\frac{dX}{X} \approx a\frac{dT_2}{T_2} + b\frac{dP_2}{P_2} + \frac{dX_c}{X_c} \Rightarrow \frac{dX_c}{X_c} = \frac{dX}{X} - a\frac{dT_2}{T_2} - b\frac{dP_2}{P_2} \Rightarrow$$

$$\frac{dX_c}{X_c} \approx \frac{dX}{X} - a\frac{d\theta}{\theta} - b\frac{d\delta}{\delta} \Rightarrow$$

$$\ln(X_c) = \ln(X) - \ln(\theta)^a - \ln(\delta)^b \Rightarrow$$

$$X_c \approx \frac{X}{\theta^a \delta^b} \quad \text{where} \quad \theta = \frac{T_2}{T_{ISA}}, \quad \delta = \frac{P_2}{P_{ISA}} \tag{2.3}$$

where

$$T_{ISA} = 518.67, \quad P_{ISA} = 14.696 \quad or \quad (T_{ISA} = 288.15, \quad P_{ISA} = 101325.353)$$

Equation (2.3) yields a familiar form and provides a simple formula from which to calculate an *approximate* standard day parameter correction. We show here θ and δ for both British units of measure as well as for SI units, however it clearly *does not matter which units are used as long as numerator and denominator have the same units*. X_c provides an approximate value for the parameter X that would be seen, under the same operating *state* but on a standard day atmospheric condition, i.e. T_{ISA} temperature and P_{ISA} pressure. Of course, the reference condition can be something other than ISA conditions, e.g., a hot tropical day, for example, if we desired to approximate, say Exhaust Gas Temperature (EGT), at a worse case condition.

The reference condition can also *refer* to an engine station (location) other than the inlet (station 2) for various applications. For example, we might be interested in correcting quantities local to an engine module, (e.g. the High Pressure Compressor, HPC, in a twin spool engine), to the engine station representing the inlet of the module under consideration, (e.g., station 2.5 for the HPC). There are various notations that are common in the industry to denote corrected parameters that refer to particular stations to which they are being corrected, however (unfortunately) there also appears that there is no universal notational convention. For example, one popular notation is to use the subscript "c" followed by the station number to denote a corrected parameter and the reference condition. If we were considering the high spool speed N_2 corrected to the inlet (station 2) we would denote its corrected form as N_{2c2}. If we were correcting this quantity to the inlet of the HPC (station 2.5), we would use the notation, N_{2c25}. Relative to Eq. (2.3), we would carry over the station identifier to the parameters θ and δ as well. For example, if we consider the classical correction for spool speed

$$N_{2c2} \approx \frac{N_2}{\sqrt{\theta_2}}, \quad N_{2c25} \approx \frac{N_2}{\sqrt{\theta_{25}}} \quad \text{where} \quad \theta_2 \equiv \frac{T_2}{T_{ISA}} \quad \text{and} \quad \theta_{25} \equiv \frac{T_{25}}{T_{ISA}} \tag{2.4}$$

Within this document, unless otherwise stated (explicitly), we will assume that the reference location is station 2, the inlet of the engine. Thus, within the context of this document, the appearance of $\boldsymbol{\theta}$ **without an explicit subscript will mean** $\boldsymbol{\theta}_{\mathbf{2}}$**.**

Thus, a corrected parameter maintains the same units and is calculated by means of theta and delta exponent corrections. We might mention at this juncture, that Eq. (2.3) is not only approximate for the reasons already alluded to, but in addition, some gas path parameters such as fuel flow (w_f), for example, are also (typically) corrected for humidity and fuel heating value. The impact of changes in viscosity with altitude (Reynolds effects) or changes in the gas composition and their impact on gas path parameters, for the most part, will not be considered in this presentation. Humidity corrections will be addressed in Chap. 8.

The values for $\boldsymbol{a}$ and $\boldsymbol{b}$ in Eq. (2.3) will, in general, vary with engine type and cycle,[2] however, there are some values which might be considered standard *common corrections* and are approximations which are commonly used in practice for all gas turbines. Chapter 3 will address these and derive many of the common gas path parameters and their classical standard day corrections.

Before leaving this topic, we might add that there are circumstances when the reference inlet condition might be something other than standard day conditions. This was alluded to above in the discussion regarding EGT. If we defined the reference condition to be a hot tropical day temperature, say 35 °C (T_{HD}), we could calculate what EGT would be at that condition ($\mathrm{EGT_{HD}}$) if we had at our disposal a measurement of EGT taken at ambient condition T_2 as follows by temperature extrapolation.

$$EGT_{HD} = \frac{EGT}{\theta_2}\theta_{HD} \quad \text{where} \quad \theta_2 \equiv \frac{T_2}{T_{ISA}} \quad \text{and} \quad \theta_{HD} \equiv \frac{T_{HD}}{T_{ISA}} \tag{2.5}$$

Next, we consider a few more (reasonable) axioms that we will assume for corrected parameters. Given the relationship in Eq. (2.3) above, it would be desirable if certain arithmetic relationships hold in *corrected parameter* space, in particular,

[2]This will be partially addressed in Chap. 7 on Empirical Methods.

that the corrected value of a sum is the sum of corrected values and likewise that the corrected value of a product is the product of corrected values. In symbols:

$$\begin{aligned} &\text{Axiom 2}: \quad Z = X + Y \Rightarrow Z_c = X_c + Y_c \\ &\text{Axiom 3}: \quad Z = XY \Rightarrow Z_c = X_c Y_c \end{aligned} \tag{2.6}$$

These are very sensible assumptions to add as axioms in that if a physical relationship exists as simple as a sum or product between gas path parameters the same relationship should hold as well at standard day conditions. By the same reasoning, we can generalize to any rational function, i.e., if g is a *rational function* (i.e., consists of sums, products and powers), then

$$\text{Axiom 4}: \quad Z = g(X_1, X_2, \ldots, X_n) \Rightarrow Z_c = g(X_{1c}, X_{2c}, \ldots, X_{nc}) \tag{2.7}$$

The axioms in Eqs. (2.6) and (2.7) represent an additional constraints since they are not directly derivable (through mathematical manipulation) from our implicit definition (Eq. (2.1)) of a corrected parameter. They are, however, reasonable constraints as far as applying it to thermodynamic gas turbine parameters. We invoke Buckingham's argument in their defense. Since physical quantities have *units* of measure, their interrelationships must be rational functions in order to maintain the suitableness and consistency of any resulting unit of measure, i.e. they must have the form of Eq. (1.2). What Eq. (2.7) (and Eq. 2.6 as a special case) dictate is that if there exists a *physical relationship* between parameter Z and parameters X_1, X_2, …, X_n, (demanded by physics) then the same interrelationship must also hold at any ambient environment and hence at any specific reference environment such as standard day conditions. This is a very reasonable assumption to add since our corrected quantities must have physical meaning beyond just an abstract mathematical formulation. Equations 2.1, 2.6, and 2.7 (combined in Eq. 2.8 below) are sufficient to allow us to derive reasonable approximations (in view of Eq. 2.3) for most gas turbine gas path parameter corrections.

$$\begin{aligned} &\text{Corrected Parameter Axioms} \\ &X = f(X_c, T_2, P_2) \\ &\text{(local Mach constant)} \\ &Z = X + Y \Rightarrow Z_c = X_c + Y_c \\ &Z = XY \Rightarrow Z_c = X_c Y_c \\ &Z = g(X_1, X_2, \ldots, X_n) \Rightarrow Z_c = g(X_{1c}, X_{2c}, \ldots, X_{nc}) \\ &\text{for any rational function } g \end{aligned} \tag{2.8}$$

Along with these basic assumptions we will make other simplifying assumptions, in particular, the assumption that ***specific heats are constant***. This will aid in the derivation of the *classical* corrections and will be stated clearly at the onset of a specific derivation when it is used. *In* Chap. 6 *this assumption will be relaxed in order to explore possible refinements in the correction.* For the remainder of this

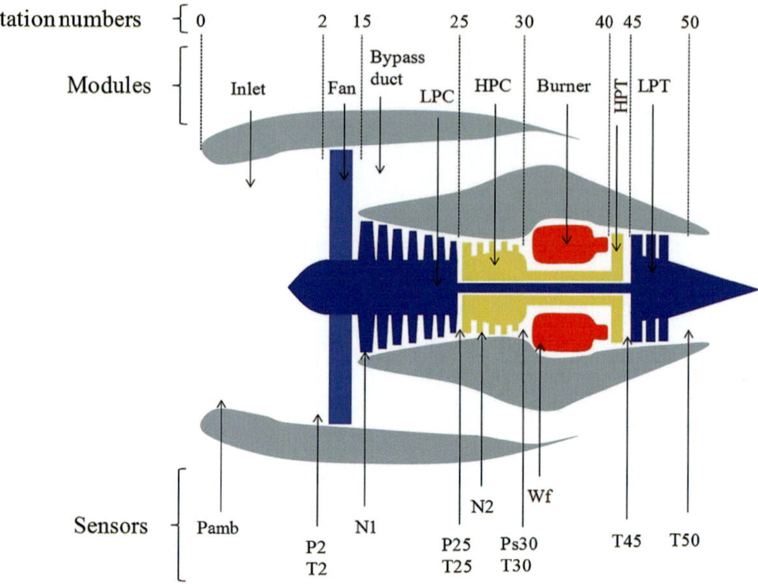

Fig. 2.1 Typical Gas Path Parameter Locations

book we will, for the most part, use a twin spool turbofan engine as the example application. For the most part, its usage will be primarily to designate engine station numbers that will be referenced in the gas path parameter symbols used throughout the text. A typical configuration[3] is depicted in Fig. 2.1 above.

Throughout this discourse, we will use the *subscript* "*c*" to denote a corrected parameter (as you see in Eq. 2.8 above).

We end this Chapter with a couple of observations that follow from the axioms that will be used in the derivations in subsequent Chapters. The first (which may be obvious to some) is that the correction of the square root of a parameter is the square root of the corrected parameter, or in symbols

$$\left(\sqrt{X}\right)_c = \sqrt{X_c} \tag{2.9}$$

This fact follows from axiom 3 through formal manipulation. Consider

$$(XY)_c = X_c Y_c \Rightarrow \left(X^2\right)_c = X_c X_c$$

Now let $Z = \sqrt{X}$ then $Z^2 = X$.
Therefore

$$X_c = \left(Z^2\right)_c = Z_c Z_c = \left(\sqrt{X}\right)_c \left(\sqrt{X}\right)_c$$

[3]Reprinted with permission from the National Aeronautics and Space Administration.

This implies that

$$\sqrt{X_c}\ \sqrt{X_c} = X_c = \left(\sqrt{X}\right)_c \left(\sqrt{X}\right)_c \Rightarrow$$
$$\frac{\sqrt{X_c}}{\left(\sqrt{X}\right)_c} = \frac{\left(\sqrt{X}\right)_c}{\sqrt{X_c}}$$
$$= \frac{1}{\sqrt{X_c}/\left(\sqrt{X}\right)_c}$$

This final equation is of the form

$$A = \frac{1}{A} \Rightarrow A^2 = 1 \Rightarrow A = \pm 1$$

Since $\sqrt{X_c}$ and $\left(\sqrt{X}\right)_c$ are, in our context, gas path physical quantities, A must be 1, thereby proving

$$\left(\sqrt{X}\right)_c = \sqrt{X_c}$$

An extension of this result would consider a general exponent power of the parameter, i.e. X^β for an arbitrary real valued exponent, β. By a similar argument, as used above for the case $\beta = 0.5$, we can show

$$\left(X^\beta\right)_c = X_c^\beta \tag{2.10}$$

We begin by considering

$$Y = X^\beta = X^{\beta-1}X \Rightarrow Y_c = \left(X^\beta\right)_c = \left(X^{\beta-1}\right)_c X_c$$

Therefore

$$\frac{X_c^\beta}{X_c^{\beta-1}} = X_c = \frac{\left(X^\beta\right)_c}{\left(X^{\beta-1}\right)_c} \Rightarrow \frac{X_c^\beta}{\left(X^\beta\right)_c} = \frac{X_c^{\beta-1}}{\left(X^{\beta-1}\right)_c}$$

Now since this last relationship must hold for arbitrary β, then these ratios must be constant, i.e.

$$\frac{X_c^\beta}{\left(X^\beta\right)_c} = const = \frac{X_c^{\beta-1}}{\left(X^{\beta-1}\right)_c}$$

Since the relationship is valid for all real β, it must be valid for $\beta = 1$ which implies that the value of $const = 1$, thereby providing the solution seen in Eq. (2.10).

Chapter 3
Common Corrections

Abstract This chapter provides the Standard Day corrections for the most common gas path parameters. These corrections are sometimes referred to as classical corrections in that, the θ and δ exponent values that are derived, are the well-recognized values that appear in the literature. The parameters considered include, Rotor Speed, Temperature, Pressure, Air Flow, Fuel Flow, Horsepower, Torque, Acceleration, Thrust, Metal Temperature Rate, Fuel Air Ratio (FAR), and Thrust Specific Fuel Consumption (TSFC). The corrections are derived in detail from the definitions, axioms, and assumptions that were proposed in Chap. 2 along with simple thermodynamic relationships relevant to the parameter under consideration. In order to avoid any circular reasoning, parameters are derived in a particular order, wherein only results from already derived parameters are utilized in a given derivation.

Keywords Classical corrections · Rotor Speed · Temperature · Pressure · Air flow · Fuel Flow · Horsepower · Torque · Acceleration · Thrust · Metal Temperature Rate · Fuel Air Ratio (FAR) · Thrust Specific Fuel Consumption (TSFC) · Theta · Delta · International Standard Atmosphere · Specific heat · Local Mach No. · Ratio of specific heats

In this Chapter, we will attempt to supply some rationale for the corrections of commonly used gas path parameters. Most will be motivated from simple thermodynamic relationships and simplifying assumptions and do not require an extensive knowledge of either thermodynamics or gas turbine operation. Where applicable we will attempt to supply the required definitions.

The classical parameter corrections for inter-stage temperatures, pressures, flow, force and power will be derived (in approximation) in the form of Eq. (2.2), i.e.,

$$X_c \approx \frac{X}{\theta^a \delta^b} \quad \text{where} \quad \theta = \frac{T_2}{T_{ISA}}, \quad \delta = \frac{P_2}{P_{ISA}}$$

using only the axioms stated in Chap. 2. Other simplifying hypotheses such as assuming ideal gas properties, constant specific heats, etc. are also invoked in the

A. J. Volponi, *Gas Turbine Parameter Corrections*,
https://doi.org/10.1007/978-3-030-41076-6_3

Table 3.1 Common gas turbine parameter corrections

Parameter	Symbol	*a*	*b*	Corrected parameter
Rotor Speed	N	0.5	0	$N_c = \frac{N}{\sqrt{\theta}}$
Airflow	w_a	−0.5	1	$w_{ac} = \frac{w_a\sqrt{\theta}}{\delta}$
Fuel Flow	w_f	0.5	1	$w_{fc} = \frac{w_f}{\theta^a\delta} = \frac{w_f}{\sqrt{\theta}\delta}$
Thrust	F_n	0	1	$F_{nc} = \frac{Fn}{\delta}$
Horse Power	HP	0.5	1	$HP_c = \frac{HP}{\sqrt{\theta}\delta}$
Torque	Q	0	1	$Q_c = \frac{Q}{\delta}$
Temperature	T	1	0	$T_c = \frac{T}{\theta^a} = \frac{T}{\theta}$
Pressure	P	0	1	$P_c = \frac{P}{\delta}$
Acceleration	$\dot{N}$	0	1	$\dot{N}_c = \frac{\dot{N}}{\delta}$
Metal Temp Rate	$\dot{T}_m$	0.74	0.8	$\dot{T}_{mc} = \frac{\dot{T}_m}{\theta^{0.74}\delta^{0.8}}$

derivations to arrive at the classical and familiar standard day corrections. Parameters, and their corrections, that are addressed in this Chapter are summarized above in Table 3.1.

The most extensively used and perhaps the simplest correction of all is the correction for temperatures, and is the first one which we shall consider.

3.1 Corrected Temperature

As a general rule, it seems reasonable that engine inlet temperature (T_2) should have a direct effect on downstream temperatures in the engine's gas path. It seems natural, for example, that for a given operating condition, increasing T_2 would in turn increase $T_{2.5}$, which in turn would increase T_3, etc. In order to motivate this concept, we need to consider the T-S diagram. To be definite, let us consider the compression induced by the Fan (inner diameter) and the Low Pressure Compressor (LPC) from station 2 to station 2.5 of our sample engine appearing (figuratively) in Fig. 3.1. Without loss of generality, an isentropic compression has been assumed. Figure 3.1 depicts a compression from P_2 to $P_{2.5}$ starting at different (inlet) temperatures (T_2 and T_2*).

At this point we will note two properties of the T-S diagram which we will state without proof. (The reader can derive these for him/herself, armed only with the elementary laws of thermodynamics.) The relationships are

1. The curves of constant pressure are monotonically increasing with entropy S
2. The constant pressure curves diverge from one another with increasing entropy S

These properties suggest that the * quantities are larger in magnitude than their non-starred counterparts and that ($T_{2.5}$ − T_2) < ($T_{2.5}$* − T_2*). What we will now show, however, is that the ratio of $T_{2.5}$ over T_2 is approximately equal to $T_{2.5}$* over T_2* !

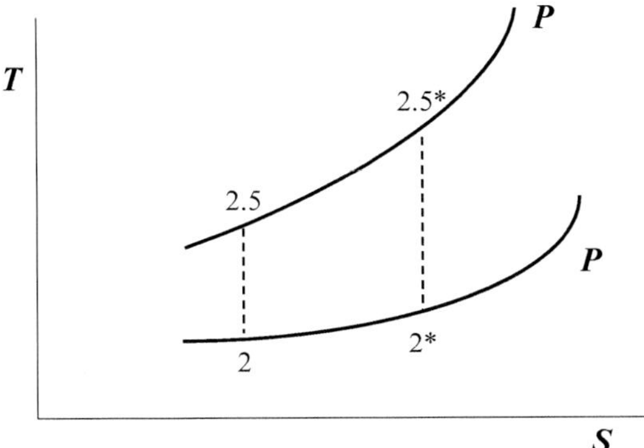

Fig. 3.1 Compression T-S diagram

To accomplish this we need one more relationship from basic thermodynamics (which follows from the definition of entropy and the perfect gas law).

$$dS = \frac{dh}{T} - R\frac{dP}{P} \tag{3.1}$$

where *S, h, T, P, R* denote entropy, enthalpy, temperature, pressure and the gas constant, respectively. Integrating Eq. (3.1) between states **2** and **2.5** and *assuming constant specific heat*, yields,

$$\begin{aligned} 0 &= \int_2^{2.5} dS = \int_2^{2.5} \frac{dh}{T} - R\int_2^{2.5} \frac{dP}{P} = cp\int_2^{2.5} \frac{dT}{T} - R\int_2^{2.5} \frac{dP}{P} \\ &= c_p \ln\left(\frac{T_{2.5}}{T_2}\right) - R\ln\left(\frac{P_{2.5}}{P_2}\right) \\ &\Rightarrow \\ \frac{T_{2.5}}{T_2} &= \left(\frac{P_{2.5}}{P_2}\right)^{R/cp} \end{aligned} \tag{3.2}$$

Likewise, integrating between 2* and 2.5* with the same assumption on specific heats, provides

$$\begin{aligned} \frac{T_{2.5}*}{T_2*} &= \left(\frac{P_{2.5}}{P_2}\right)^{R/cp} \\ &= \frac{T_{2.5}}{T_2} \\ &\Rightarrow \\ T_{2.5}* &= \frac{T_{2.5}}{(T_2/T_2*)} = \frac{T_{2.5}}{\theta} \end{aligned} \tag{3.3}$$

as required. By using the same arguments from station 2.5 to 3, we have that

$$\frac{T_3*}{T_{2.5}*} = \frac{T_3}{T_{2.5}} \Rightarrow \frac{T_3*}{T_3} = \frac{T_{2.5}*}{T_{2.5}} = \frac{T_2*}{T_2} \Rightarrow T_3* = \frac{T_3}{(T_2/T_2*)} = \frac{T_3}{\theta} \tag{3.4}$$

and so forth. Therefore, we have in general

$$T_c = \frac{T}{\theta} \tag{3.5}$$

The next most common gas path parameter that immediately comes to mind is pressure, which we consider next.

3.2 Corrected Pressure

The pressure changes experienced at various stations throughout the engine's gas-path are the effect of either compressions or expansions resulting from the action of the engine's turbomachinery. For purposes of motivating the correction for pressure, we can consider without loss of generality, a compression process; the argument for expansion pressures would be similar. Let us consider, for example, the pressure $P_{2.5}$ at the exit of the LPC. This pressure is related to temperature and pressure by virtue of the following relationship

$$\frac{T_{2.5}}{T_2} = \left(\frac{P_{2.5}}{P_2}\right)^{\frac{\gamma-1}{\gamma\eta}} \quad \text{where } \eta = \text{polytropic efficiency} \quad \gamma = cp/cv = \text{ratio of specific heats} \tag{3.6}$$

Thus, since $T_{2.5}/\theta$ = constant implies $T_{2.5}/T_2$ = constant, we have that $P_{2.5}/P_2$ = constant, which in turn implies that $P_{2.5}/\delta$ is constant, which establishes the correction at station 2.5. Note, that implicit in this argument is that the ratio of specific heats γ and efficiency η do not change. Moving downstream to station 3 we have that

$$\frac{P_3}{P_{2.5}} = \left(\frac{T_3}{T_{2.5}}\right)^{\frac{\gamma\eta}{\gamma-1}} = \left(\frac{T_3/\theta}{T_{2.5}/\theta}\right)^{\frac{\gamma\eta}{\gamma-1}} = const.$$

therefore

$$\frac{P_3}{\delta} = \frac{P_3}{P_{2.5}}\frac{P_{2.5}}{\delta} = const \tag{3.7}$$

Thus, we proceed downstream and establish in general that P/δ = constant as required.

$$P_c = \frac{P}{\delta} \tag{3.8}$$

3.3 Corrected Air Flow

Consider air flow w_a (in g/s) moving with uniform velocity v through a pipe having cross chapteral area A (m^2). Then

$$w_a = \rho A v = \text{density}\left(\text{g/m}^3\right) \times \text{area}\left(\text{m}^2\right) \times \text{velocity}(\text{m/s}) \tag{3.9}$$

For an ideal gas, we also know that $P_s = \rho RT$, from which it follows that

$$\frac{w_a}{A} = \rho v = \frac{P_s}{RT_s} v \tag{3.10}$$

where the subscripts s and t denote static and total quantities respectively. From the definition of Mach number M_n

$$M_n = v/\sqrt{\gamma RT_s} \tag{3.11}$$

we obtain the relationship

$$\frac{w_a}{A} = \frac{P_s M_n \sqrt{\gamma RT_s}}{RT_s} \quad \Rightarrow \quad \frac{w_a \sqrt{T_t}}{AP_t} = \left(\frac{P_s}{P_t}\right)\sqrt{\frac{T_t}{T_s}} M_n \sqrt{\frac{\gamma}{R}} \tag{3.12}$$

We know, however, that for an isentropic process

$$\frac{T_t}{T_s} = 1 + \frac{\gamma - 1}{2} M_n{}^2 \quad \text{and} \quad \frac{P_t}{P_s} = \left(1 + \frac{\gamma - 1}{2} M_n{}^2\right)^{\frac{\gamma}{\gamma - 1}} \tag{3.13}$$

Substituting these terms into Eq. (3.12), we obtain

$$\frac{w_a \sqrt{T_t}}{AP_t} = \sqrt{\frac{\gamma}{R}} M_n \left(1 + \frac{\gamma - 1}{2} M_n{}^2\right)^{\frac{\gamma + 1}{2(1 - \gamma)}} \tag{3.14}$$

Multiplying both sides of Eq. (3.14) by $(\theta/T_t)^{1/2}/(\delta/P_t)$ to get the proper units (g/s) yields

$$\frac{w_a \sqrt{\theta}}{\delta} = \frac{A}{(P_t/\delta)} \sqrt{\frac{\gamma (T_t/\theta)}{R}} M_n \left(1 + \frac{\gamma - 1}{2} M_n{}^2\right)^{\frac{\gamma + 1}{2(1 - \gamma)}} = \text{constant} \tag{3.15}$$

for M_n = constant, thus providing a *corrected* quantity.[1]

$$w_{ac} = \frac{w_a \sqrt{\theta}}{\delta} \tag{3.16}$$

[1] T/θ and P/δ are corrected parameters and hence constant by virtue of the previous derivations.

3.4 Corrected Fuel Flow

Consider the (simplified) energy equation bounding the combustor

$$\frac{BTU}{\sec} = \eta_b q w_f = (w_a + w_f)\Delta h = (w_a + w_f)(h_5 - h_4) \tag{3.17}$$

where h denotes enthalpy, q denotes the heating value of the fuel and η denotes the adiabatic efficiency of the combustor. If we assume that (at a given operating point) efficiency and fuel heating value are constant, then if we take logs of both sides of Eq. (3.17) and differentiate, we obtain

$$\begin{aligned}\frac{dw_f}{w_f} &= \frac{d(w_a + w_f)}{(w_a + w_f)} + \frac{d(\Delta h)}{\Delta h}\\ &= \left(\frac{w_a}{(w_a + w_f)}\right)\frac{dw_a}{w_a} + \left(\frac{w_f}{(w_a + w_f)}\right)\frac{dw_f}{w_f}\\ &\quad + \left(\frac{cp_4 T_4}{\Delta h}\right)\frac{dT_4}{T_4} - \left(\frac{cp_3 T_3}{\Delta h}\right)\frac{dT_3}{T_3}\end{aligned} \tag{3.18}$$

A little algebraic manipulation on Eq. (3.18) gives

$$\left(\frac{w_a}{w_a + w_f}\right)\frac{dw_f}{w_f} = \left(\frac{w_a}{(w_a + w_f)}\right)\frac{dw_a}{w_a} + \left(\frac{cp_4 T_4}{\Delta h}\right)\frac{dT_4}{T_4} - \left(\frac{cp_3 T_4}{\Delta h}\right)\frac{dT_3}{T_3} \tag{3.19}$$

Now, at a given operating point, we can assume that the following relationships hold:

$$\begin{aligned} & & \frac{w_a\sqrt{\theta}}{\delta} &= const & &\Rightarrow & \frac{dw_a}{w_a} + 1/2\frac{dT_3}{T_3} - \frac{dP_3}{P_3} &= 0\\ \frac{T_3}{\theta} & \text{ and } & \frac{T_4}{\theta} &= const & &\Rightarrow & \frac{dT_3}{T_3} = \frac{dT_2}{T_2} &= \frac{dT_4}{T_4}\\ & & \frac{P_3}{\delta} &= const & &\Rightarrow & \frac{dP_3}{P_3} &= \frac{dP_2}{P_2}\end{aligned} \tag{3.20}$$

Thus Eq. (3.19) can be re-written as,

$$\begin{aligned}\frac{dw_f}{w_f} &= \frac{dP_2}{P_2} - 1/2\frac{dT_2}{T_2} + \left(\frac{w_a + w_f}{w_a}\frac{cp_4 T_4}{\Delta h}\right)\frac{dT_2}{T_2} - \left(\frac{w_a + w_f}{w_a}\frac{cp_3 T_3}{\Delta h}\right)\frac{dT_2}{T_2}\\ &= \frac{dP_2}{P_2} + \underbrace{\left[\left(1 + \frac{w_f}{w_a}\right)\left(\frac{cp_4 T_4}{\Delta h} - \frac{cp_3 T_3}{\Delta h}\right) - 1/2\right]}_{x}\frac{dT_2}{T_2}\\ &= \frac{dP_2}{P_2} + x\frac{dT_2}{T_2}\\ &\Rightarrow \frac{w_f}{\delta\theta^x} = const\end{aligned} \tag{3.21}$$

Furthermore, if we make the assumption that specific heats are constant, i.e. $cp_3 = cp_4$ then the exponent x reduces to 1/2 + Fuel-Air Ratio = (approximately) 1/2 and we obtain the classical normalization

$$w_{fc} = \frac{w_f}{\delta\ \sqrt{\theta}} \tag{3.22}$$

3.5 Corrected Net Thrust

Expressions for the net thrust produced by a gas turbine engine will depend upon the configuration of the engine, i.e., whether it is a turbojet or a turbofan, mixed flow or non-mixed flow, as well as an expression for ram drag as a function of Mach number. To keep the derivation as simple as possible, we will assume a static thrust expression, i.e., Mach = 0, no ram drag and, we will assume for simplicity an engine configuration which is that of a *mixed flow turbofan*. It should be noted that derivation that follows can be repeated (in spirit) with more complicated expressions for net thrust for other engine types and flight conditions to arrive at similar representations for corrected thrust. The thrust produced by this engine depends on the momentum of the air ejected from the exhaust nozzle as well as the net force (due to pressure) acting across the total area of the nozzle. In symbols

$$F_n = const\ w_{a9} v_9 + A_9(P_9 - P_{s9}) \tag{3.23}$$

where

$$\begin{aligned} \text{station } 8 &= \text{Mixing plane in the exhaust} \\ \text{station } 9 &= \text{Engine exhaust exit plane} \\ w_{a9} = w_{a8} &= \text{.Total airflow at nozzle} \\ A_9 &= \text{.Nozzle Area} \\ P_9 &= \text{.Nozzle exit total pressure} \\ P_{s9} = P_2 &= \text{.Nozzle exit static pressure (ambient)} \\ v_9 = \sqrt{const\,(h_9 - h_{s9})} &= \text{.Velocity at exhaust nozzle} \\ h_9 &= \text{.Total (stagnation) enthalpy at exit} \\ h_{s9} &= \text{.Static enthalpy at exit} \end{aligned}$$

The expression for the exit velocity follows directly from the definition of stagnation (total) enthalpy. Similarly, there is an expression for the velocity at station 8 (mixer) involving total and static enthalpies at station 8. The flow from station 8 to 9 (through a nozzle) we will assume to be adiabatic, i.e., no work done, no heat added. By conservation of energy, the total enthalpies are equal ($h_8 = h_9$). If we denote the change in enthalpy $(h_9 - h_{s9}) = (h_8 - h_{s9}) = \Delta h$, then we may rewrite Eq. (3.23) as

$$F_n = const\ \ w_{a8} v_8 \sqrt{const\ \ \Delta h} + A_9(P_9 - P_{s9}) \tag{3.24}$$

Taking logs of both sides of Eq. (3.24) and differentiating, produces

$$\begin{aligned}
\frac{dF_n}{F_n} &= \frac{(const\, w_{a8})\sqrt{const\, \Delta h}}{F_n}\left[\frac{dw_{a8}}{w_{a8}} + \frac{1}{2}\frac{d\Delta h}{\Delta h}\right] \\
&\quad + \frac{A_9(P_9 - P_{s9})}{F_n}\left[\frac{dA_9}{A_9} + \left(\frac{P_9}{P_9 - P_2}\right)\frac{dP_9}{P_9} - \left(\frac{P_2}{P_9 - P_2}\right)\frac{dP_2}{P_2}\right] \\
&= \frac{(const\, w_{a8})\sqrt{const\, \Delta h}}{F_n}\left[\frac{dw_{a8}}{w_{a8}} + \left(\frac{h_8}{2\Delta h}\right)\frac{dh_8}{h_8} - \left(\frac{h_{s9}}{2\Delta h}\right)\frac{dh_{s9}}{h_{s9}}\right] \\
&\quad + \frac{A_9(P_9 - P_{s9})}{F_n}\left[\underbrace{\frac{dA_9}{A_9}}_{\substack{\text{zero} \\ \text{(fixed geometry)}}} + \left(\frac{P_9}{P_9 - P_2}\right)\frac{dP_9}{P_9} - \left(\frac{P_2}{P_9 - P_2}\right)\frac{dP_2}{P_2}\right]
\end{aligned}$$

$$\begin{aligned}
\frac{dF_n}{F_n} &= \frac{(const\, w_{a8})\sqrt{const\, \Delta h}}{F_n}\left[\frac{dw_{a8}}{w_{a8}} + \left(\frac{cp_8 T_8}{2\Delta h}\right)\frac{dT_8}{T_8} - \left(\frac{cp_9 T_{s9}}{2\Delta h}\right)\frac{dT_{s9}}{T_{s9}}\right] \\
&\quad + \frac{A_9(P_9 - P_{s9})}{F_n}\left[\left(\frac{P_9}{P_9 - P_2}\right)\frac{dP_9}{P_9} - \left(\frac{P_2}{P_9 - P_2}\right)\frac{dP_2}{P_2}\right]
\end{aligned} \tag{3.25}$$

Now, at a given engine operating point (steady state), the following conditions hold

$$\begin{aligned}
\frac{w_{a8}\sqrt{\theta_8}}{\delta_8} &= \text{constant} \quad \Rightarrow \quad \frac{dw_{a8}}{w_{a8}} + \frac{1}{2}\frac{dT_8}{T_8} - \frac{dP_8}{P_8} = 0 \\
\frac{T_8}{\theta_2} &= \text{constant} \quad \Rightarrow \quad \frac{dT_8}{T_8} - \frac{dT_2}{T_2} = 0 \\
\frac{T_{s9}}{\theta_2} &= \text{constant} \quad \Rightarrow \quad \frac{dT_{s9}}{T_{s9}} - \frac{dT_2}{T_2} = 0 \\
\frac{P_8}{\delta_2} &= \text{constant} \quad \Rightarrow \quad \frac{dP_8}{P_8} - \frac{dP_2}{P_2} = 0 \\
\frac{P_9}{\delta_2} &= \text{constant} \quad \Rightarrow \quad \frac{dP_9}{P_9} - \frac{dP_2}{P_2} = 0
\end{aligned}$$

Substituting these values into Eq. (3.25), we obtain

$$\frac{dF_n}{F_n} = \frac{(const\, w_{a8})\sqrt{const\, \Delta h}}{F_n}\left[\frac{dP_8}{P_8} + \underbrace{\left(\frac{cp_8 T_8}{2\Delta h} - \frac{cp_9 T_{s9}}{2\Delta h} - \frac{1}{2}\right)}_{x}\frac{dT_2}{T_2}\right]$$
$$+ \frac{A_9(P_9 - P_{s9})}{F_n}\frac{dP_2}{P_2} \tag{3.26}$$
$$= \underbrace{\frac{(const\, w_{a8})\sqrt{const\, \Delta h}}{F_n}}_{y_1}\left[\frac{dP_8}{P_8} + x\frac{dT_2}{T_2}\right] + \underbrace{\frac{A_9(P_9 - P_{s9})}{F_n}}_{y_2}\frac{dP_2}{P_2}$$

This yields

$$\frac{dF_n}{F_n} = (y_1 + y_2)\frac{dP_2}{P_2} + y_1 x\frac{dT_2}{T_2} = \frac{dP_2}{P_2} + y_1 x\frac{dT_2}{T_2}$$
$$\Rightarrow \frac{F_n}{\delta_2 {\theta_2}^{y_1 x}} = \text{constant} \tag{3.27}$$

Since the specific heats at station 8 and 9 are approximately equal (since $T_8 \approx T_{9s}$), we may assume a common specific heat cp. In this case, the theta exponent x reduces to 0 (since $\Delta h = cp\Delta T$). Thus, we arrive at the classical correction for thrust, namely

$$F_{nc} = \frac{F_n}{\delta} \tag{3.29}$$

3.6 Corrected Rotational Speeds

Tangential velocity is related to rotational speed (rpm) by the radius of the object (compressor or turbine blade) in question (see Fig. 3.2). It is also related to acoustic velocity by Mach number and the square root of temperature.

Mathematically, we have

$$v = rN = M_n\sqrt{\gamma RT} \tag{3.30}$$

where v is the velocity in m/s, r is the radius (m), N is the rotational speed (rad/s), γ and R are constants (ratio of specific heats and the gas constant respectively), and M_n is Mach number. Taking logs and differentiating Eq. (3.30), we obtain

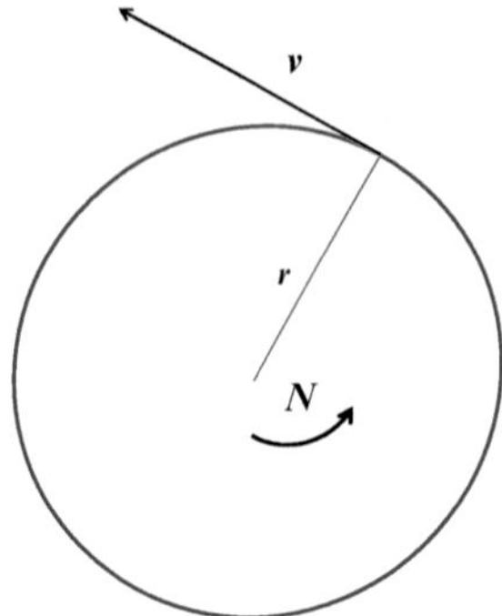

Fig. 3.2 Rotational speed

$$\frac{dN}{N} = \frac{dM_n}{M_n} + \frac{1}{2}\frac{dT}{T} = \frac{1}{2}\frac{dT}{T} \quad \text{and since Mach} = \text{constant}$$
$$\Rightarrow \frac{dN}{N} - \frac{1}{2}\frac{d\theta}{\theta} = 0$$
$$\Rightarrow \frac{N}{\sqrt{\theta}} = \text{constant}$$

Therefore

$$N_c = \frac{N}{\sqrt{\theta}} \tag{3.31}$$

3.7 Corrected Horsepower

Horsepower required (or developed) by a compressor (or turbine) is given by

$$\begin{aligned} &HP \propto (w_a \Delta h) \Rightarrow HP = const(w_a \Delta h) \\ &\therefore \quad HP = const(w_a cp \Delta T) \end{aligned} \tag{3.32}$$

Invoking axiom 3, this relationship must also hold for corrected quantities, i.e.

$$\begin{aligned} HP_c &= const \;\; w_{ac} cp \Delta T_c \\ &= const \frac{w_a \sqrt{\theta}}{\delta} cp \frac{\Delta T}{\theta} \\ &= \frac{const\, w_a cp \Delta T}{\delta \sqrt{\theta}} \end{aligned}$$

Therefore

$$HP_c = \frac{HP}{\delta \sqrt{\theta}} \tag{3.33}$$

3.8 Corrected Torque

The correction for torque follows from corrections for speed and horsepower, since horesepower is the product of torque and speed. It follows (from axiom 3) that a corrected horsepower should be a product of corrected torque and corrected speed, i.e.

$$\begin{aligned} HP_c &= Q_c N_c \\ &= \frac{HP}{\delta\sqrt{\theta}} = Q_c \frac{N}{\sqrt{\theta}} \\ &\Rightarrow \\ Q_c &= \frac{HP}{N\delta} \end{aligned}$$

Thus,

$$Q_c = \frac{Q}{\delta} \tag{3.34}$$

3.9 Corrected Acceleration

From Newton's Law, Torque (force) = Moment of Inertia (I) × Acceleration. In symbols

$$Q = I\dot{N} \tag{3.35}$$

Therefore by axiom 3, we must have

$$Q_c = I\dot{N}_c \tag{3.36}$$

This implies

$$\begin{aligned} \frac{Q}{\delta} &= I\dot{N}_c \\ &= I\frac{\dot{N}}{\delta} \Rightarrow \\ \dot{N}_c &= \frac{\dot{N}}{\delta} \end{aligned} \tag{3.37}$$

3.10 Corrected Metal Temperature Rate

The (time) rate of change of a metal temperature is typically modeled as a first order lag heat transfer between the gas path temperature T and the metal temperature T_m. In symbols,

$$\dot{T}_m = \frac{1}{\tau}(T - T_m)$$
$$\text{where time constant } \tau = \frac{mc_p}{HA} \tag{3.38}$$

The McAdams correlation for turbulent flow over a flat plate [7] is given by

$$\text{Nu} = 0.023(\text{ Re })^{0.8}(\text{Pr})^{0.4} \tag{3.39}$$

where Nu, Re, and Pr are the Nusselt's, Reynold's and Prandtl's numbers respectively which are defined as follows:

$$\text{Nu} = \frac{HD}{\kappa}, \quad \text{Re} = \frac{\rho v D}{\mu}, \quad \text{Pr} = \frac{cp\mu}{\kappa} \tag{3.40}$$

Substituting and re-arranging terms we can obtain the following form:

$$\begin{aligned} H &= 0.023\frac{\kappa}{D}\left(\frac{\rho v D}{\mu}\right)^{0.8}\left(\frac{cp\,\mu}{\kappa}\right)^{0.4} \\ &= 0.023\frac{\kappa^{0.6}cp^{0.4}}{D^{0.2}}\frac{\rho^{0.8}v^{0.8}}{\mu^{0.4}} \\ &= \left(\frac{0.023\kappa^{0.6}}{cp^{0.6}\mu^{0.6}}\right)\frac{cp\mu^{0.2}\rho^{0.8}v^{0.8}}{D^{0.2}} \\ &= \left(\frac{0.023}{\text{Pr}^{0.6}}\right)\frac{cp\mu^{0.2}\rho^{0.8}v^{0.8}}{D^{0.2}} \\ &\approx 0.027\frac{cp\mu^{0.2}\rho^{0.8}v^{0.8}}{D^{0.2}} \end{aligned} \tag{3.41}$$

where the final approximation is obtained by noting that Prandtl's number is approximately constant in the range of interest [7]. At this juncture we may recall that mass flow $w_a = \rho A v$ was corrected by square root of theta over delta. Combining these observations, we have

$$\frac{w_a}{A} = \frac{w_{ac}\delta}{\sqrt{\theta}} = \rho v \tag{3.42}$$

Therefore,

$$(\rho v)^{0.8} = (w_{ac})^{0.8}\delta^{0.8}\theta^{-0.4} \tag{3.43}$$

Using the approximation at the end of Eq. (3.41) to define the reference condition H^* and noting that $D = D^*$, we can write the ratio

$$\begin{aligned}\frac{H*}{H} &= \left(\frac{cp*}{cp}\right)\left(\frac{\mu*}{\mu}\right)^{0.2}\left(\frac{\rho*v*}{\rho v}\right)^{0.8}\\ &= \left(\frac{cp*}{cp}\right)\left(\frac{1}{\theta^n}\right)^{0.2}\left(\frac{\sqrt{\theta}}{\delta}\right)^{0.8}\\ &= \left(\frac{cp*}{cp}\right)\theta^{0.4-0.2n}\delta^{-0.8}\end{aligned} \tag{3.44}$$

where we have assumed that the ratio μ*/μ is solely a function of temperature (θ). Now using Eq. (3.38) to define $\dot{T}_m* = \dot{T}_{mc}$ in a similar manner, we can form the appropriate ratio and use Eq. (3.44) to simplify the expression.

$$\begin{aligned}\frac{\dot{T}_{m\,c}}{\dot{T}_m} &= \left(\frac{H*}{H}\right)\left(\frac{cp}{cp*}\right)\frac{T_c - T_{m\,c}}{T - T_m}\\ &= \theta^{0.4-0.2n}\delta^{-0.8}\frac{\left(T/\theta\right) - \left(T_m/\theta\right)}{T - T_m}\\ &= \theta^{-0.6-0.2n}\delta^{-0.8}\\ &\Rightarrow\\ \dot{T}_{m\,c} &= \frac{\dot{T}_m}{\theta^{0.6+0.2n}\delta^{0.8}}\end{aligned} \tag{3.45}$$

The value of n appearing in the theta exponent can be determined experimentally by correlating viscosity with temperature from which it is observed that n varies from 0.8 at low temperature to 0.6 at high temperature. These values provide a theta exponent ranging from 0.76 to 0.72, respectively. The average value of 0.74 is recommended, thus providing the requisite correction.

$$\dot{T}_{mc} = \frac{\dot{T}_m}{\theta^{0.74}\delta^{0.8}} \tag{3.46}$$

3.11 Corrected Fuel Air Ratio

This follows simply from definition and axiom 3, i.e.,

$$FAR = \frac{w_f}{w_a} \tag{3.47}$$

and

$$FAR_c = \frac{w_{fc}}{w_{ac}} = \frac{\frac{w_f}{\delta\sqrt{\theta}}}{\frac{w_a\sqrt{\theta}}{\delta}} = \frac{\frac{w_f}{w_a}}{\theta} \tag{3.48}$$

Thus

$$FAR_c = \frac{FAR}{\theta} \tag{3.49}$$

3.12 Thrust Specific Fuel Consumption

Likewise for TSFC, we have

$$TSFC = \frac{w_f}{F_n} \Rightarrow \frac{TSFC}{\sqrt{\theta}} = \frac{\frac{w_f}{\delta\sqrt{\theta}}}{\frac{F_n}{\delta}} = TSFC_c \tag{3.50}$$

$$TSFC_c = \frac{TSFC}{\sqrt{\theta}} \tag{3.51}$$

This concludes the derivation of the classical corrections for the common gas path parameters contained in Table 3.1. We will return to these in Chaps. 6 and 7 to explore some refinements to the theta and delta exponents.

In the next Chapter we consider the corrections to be applied to time derivatives of these common parameters beyond acceleration and metal temperature, which were derived above.

Chapter 4
Time Derivative Corrections

Abstract This chapter provides the Standard Day corrections for the time rate of change, (i.e., time derivative) for the gas path parameters encountered in Chap. 3. The parameters considered include, Acceleration, Temperature, Pressure, Air Flow, Fuel Flow, Horsepower, Torque, Thrust, Fuel Air Ratio (FAR), and Thrust Specific Fuel Consumption (TSFC). As in the previous chapter, corrections are derived in detail from the definitions, axioms, and assumptions that were proposed in Chap. 2 along with simple thermodynamic relationships relevant to the parameter under consideration. In order to avoid any circular reasoning, parameters are derived in a particular order, wherein only results from already derived parameters are utilized in a given derivation.

Keywords Time-derivative corrections · Rate of change · Temperature · Pressure · Air Flow · Fuel Flow · Horsepower · Torque · Acceleration · Thrust · Fuel Air Ratio (FAR) · Thrust Specific Fuel Consumption (TSFC) · Theta · Delta · Specific heat · Local Mach No. · Ratio of specific heats

In this chapter we will extend the classical parameter corrections to include the correction of the time derivatives of a gas turbine parameter. Again, we will rely only on our axioms from Chap. 2 and basic thermodynamic relationships. The astute reader may have already observed, from our classical correction results from the preceding Chapter, that the corrected parameter of a time derivative is *not* equal to the time derivative of the corrected parameter, i.e.,

$$\dot{X}_c \neq \frac{d}{dt}(X_c)$$

The familiar *counter-example* is the time derivative of (spool) speed, i.e. acceleration. We already know from Table 3.1 that

A. J. Volponi, *Gas Turbine Parameter Corrections*,
https://doi.org/10.1007/978-3-030-41076-6_4

$$\dot{N}_c = \frac{\dot{N}}{\delta} \neq \frac{\dot{N}}{\sqrt{\theta}} = \frac{d}{dt}\left(\frac{N}{\sqrt{\theta}}\right) = \frac{d}{dt}(N_c)$$

Thus, we cannot simply say, for example, that

$$\ddot{N}_c = \frac{d}{dt}(\dot{N}_c) = \frac{d}{dt}\left(\frac{\dot{N}}{\delta}\right) = \frac{\ddot{N}}{\delta}\,(\text{false})$$

In fact, we will demonstrate that the true relationship for the time derivative of acceleration (under the assumptions of our axioms) is as follows:

$$\ddot{N}_c = \frac{\ddot{N}\sqrt{\theta}}{\delta^2}$$

The derivation is a bit involved and not as direct as those encountered in Chap. 3. In fact we will first show that the time derivative correction for acceleration is closely linked to the time derivative correction for horsepower. More precisely, we will show that

$$\dot{HP}_c = \frac{\dot{HP}}{\delta^2} \quad \text{if and only if} \quad \ddot{N}_c = \frac{\ddot{N}\sqrt{\theta}}{\delta^2}$$

4.1 Correction of Horsepower and Acceleration Rate of Change

We begin with the definition of horsepower as the product of torque and rotational speed and Newton's law, i.e.

$$HP = QN = I\dot{N}N \tag{4.1}$$

where I represents the moment of inertia. Differentiating Eq. (4.1) (with respect to time) and applying the derivative of a product form, we obtain

$$\dot{HP} = I(\ddot{N}N + \dot{N}\dot{N}) = I\left(\ddot{N}N + \dot{N}^2\right) \tag{4.2}$$

Invoking axiom 2 and 3 for a sum and a product, (which will be used repeatedly in the course of this chapter), the relationship must hold at a standard day conditions, namely that

$$\dot{HP}_c = I\left(\ddot{N}_c N_c + \dot{N}_c^2\right) \approx I\left(\ddot{N}_c \frac{N}{\sqrt{\theta}} + \left(\frac{\dot{N}}{\delta}\right)^2\right) \tag{4.3}$$

Dividing the Eq. (4.2) by δ^2 and subtracting it from the Eq. (4.3), yields

$$\left(\dot{HP}_c - \frac{\dot{HP}}{\delta^2}\right) \approx I\left(\ddot{N}_c \frac{N}{\sqrt{\theta}} + \left(\frac{\dot{N}}{\delta}\right)^2 - \frac{\ddot{N}N}{\delta^2} - \frac{\dot{N}^2}{\delta^2}\right) \Rightarrow$$
$$\left(\dot{HP}_c - \frac{\dot{HP}}{\delta^2}\right) \approx I\left(\ddot{N}_c \frac{N}{\sqrt{\theta}} - \frac{\ddot{N}N}{\delta^2}\right) = I\left(\ddot{N}_c \frac{N}{\sqrt{\theta}} - \frac{\ddot{N}\sqrt{\theta}}{\delta^2}\frac{N}{\sqrt{\theta}}\right) \Rightarrow$$
$$\left(\dot{HP}_c - \frac{\dot{HP}}{\delta^2}\right) \approx I\frac{N}{\sqrt{\theta}}\left(\ddot{N}_c - \frac{\ddot{N}\sqrt{\theta}}{\delta^2}\right) \Rightarrow$$

$$\left(\dot{HP}_c - \frac{\dot{HP}}{\delta^2}\right) \approx IN_c\left(\ddot{N}_c - \frac{\ddot{N}\sqrt{\theta}}{\delta^2}\right) \tag{4.4}$$

Clearly, one solution to Eq. (4.4) is for both sides to equal zero which would yield the proposition

$$\dot{HP}_c = \frac{\dot{HP}}{\delta^2} \quad \text{if and only if} \quad \ddot{N}_c = \frac{\ddot{N}\sqrt{\theta}}{\delta^2}$$

Furthermore, since Eq. (4.4) must hold for any ambient condition, i.e. for all θ and δ. We also have that

$$\dot{HP}_c = const_1$$
$$IN_c = const_2$$
$$\ddot{N}_c = const_3$$

since they are corrected quantities. Therefore we can re-write Eq. (4.4) as

$$\left(const_1 - \frac{\dot{HP}}{\delta^2}\right) = const_2\left(const_3 - \frac{\ddot{N}\sqrt{\theta}}{\delta^2}\right) \Rightarrow$$
$$\frac{\dot{HP}}{\delta^2} = const_1 - const_2\,const_3 + const_2\frac{\ddot{N}\sqrt{\theta}}{\delta^2} \tag{4.5}$$
$$\frac{\dot{HP}}{\delta^2} = const_4 + const_2\frac{\ddot{N}\sqrt{\theta}}{\delta^2} \quad \text{for all } \theta, \delta$$

Thus for any δ and any two thetas, $\theta^{(1)}$ and $\theta^{(2)}$ we must have

$$\frac{\dot{HP}}{\delta^2} = const_4 + const_2\frac{\ddot{N}\sqrt{\theta^{(1)}}}{\delta^2} \quad \text{and}$$
$$\frac{\dot{HP}}{\delta^2} = const_4 + const_2\frac{\ddot{N}\sqrt{\theta^{(2)}}}{\delta^2} \tag{4.6}$$

Subtracting these two expressions from each other yields

$$0 = const_2 \left(\frac{\ddot{N}\sqrt{\theta^{(1)}}}{\delta^2} - \frac{\ddot{N}\sqrt{\theta^{(2)}}}{\delta^2} \right) \tag{4.7}$$

Since Eq. (4.7) must hold for *arbitrary* δ and *any* two thetas, $\theta^{(1)}$ and $\theta^{(2)}$, we have that

$$\frac{\ddot{N}\sqrt{\theta^{(1)}}}{\delta^2} = \frac{\ddot{N}\sqrt{\theta^{(2)}}}{\delta^2} = const \Rightarrow \frac{\ddot{N}\sqrt{\theta}}{\delta^2} = const \quad \text{for all } \theta, \delta \tag{4.8}$$

Therefore we have our time derivative of acceleration correction.

$$\ddot{N}_c = \frac{\ddot{N}\sqrt{\theta}}{\delta^2} \tag{4.9}$$

Invoking Eq. (4.5), we have the time derivative correction for *HP*, namely

$$\dot{H}P_c = \frac{\dot{H}P}{\delta^2} \tag{4.10}$$

4.2 Correction of Torque Rate of Change

Since $HP = QN$, we have by Eq. (4.10)

$$HP_c = \frac{HP}{\sqrt{\theta}\delta} = \frac{QN}{\sqrt{\theta}\delta} = \frac{Q}{\delta}\frac{N}{\sqrt{\theta}} = Q_c N_c \tag{4.11}$$

Differentiating with respect to time, we obtain

$$\dot{H}P = \dot{N}Q + N\dot{Q} \Rightarrow \dot{H}P_c = \dot{N}_c Q_c + N_c \dot{Q}_c \tag{4.12}$$

implying that

$$\frac{\dot{H}P}{\delta^2} = \frac{N\dot{Q}}{\delta^2} + \frac{\dot{N}Q}{\delta^2} = \frac{N}{\sqrt{\theta}}\frac{\dot{Q}\sqrt{\theta}}{\delta^2} + \frac{\dot{N}}{\delta}\frac{Q}{\delta} = N_c\frac{\dot{Q}\sqrt{\theta}}{\delta^2} + \dot{N}_c Q_c \tag{4.13}$$

Now subtracting Eq. (4.13) from Eq. (4.12), we obtain, yields

$$\left(\dot{H}P_c - \frac{\dot{H}P}{\delta^2}\right) = \left(\dot{Q}_c - \frac{\dot{Q}\sqrt{\theta}}{\delta^2}\right)N_c \;\Rightarrow\; 0 = \left(\dot{Q}_c - \frac{\dot{Q}\sqrt{\theta}}{\delta^2}\right)N_c$$
$$\Rightarrow \left(\dot{Q}_c - \frac{\dot{Q}\sqrt{\theta}}{\delta^2}\right) = 0 \tag{4.14}$$

This provides the requisite correction:

$$\dot{Q}_c = \frac{\dot{Q}\sqrt{\theta}}{\delta^2} \tag{4.15}$$

4.3 Correction of Temperature Rate of Change

Since tangential velocity v is related to rotational speed by the radius of the object, r, we have that

$$v = rN = M_n\sqrt{\gamma R T_s} \tag{4.16}$$

If we assume local Mach number and ratio of specific heats are constant, then since

$$\frac{T}{T_s} = 1 + \frac{\gamma - 1}{2}M_n^2 \approx const \Rightarrow rN = M_n\sqrt{\gamma R T_s} = M_n\sqrt{\gamma\ R\ const\ T}$$
$$= const\sqrt{T} \tag{4.17}$$

Differentiating Eq. (4.17) with respect to time, we obtain

$$\dot{N} = const\frac{1}{\sqrt{T}}\ \dot{T} \Rightarrow$$
$$\frac{\dot{N}}{\delta} = const\frac{1}{\sqrt{T/\theta}}\ \frac{\dot{T}}{\delta\sqrt{\theta}} = const\frac{1}{\sqrt{T_c}}\ \frac{\dot{T}}{\delta\sqrt{\theta}} \tag{4.18}$$

But, the first expression in Eq. (4.18) must also hold for corrected parameters by our axiom assumptions,[1] therefore

$$\dot{N}_c = const\frac{1}{\sqrt{T_c}}\dot{T}_c \Rightarrow$$
$$\frac{\dot{N}}{\delta} = \dot{N}_c = const\frac{1}{\sqrt{T_c}}\dot{T}_c \tag{4.19}$$

[1]The correction of a square root is the square root of the correction : see Eq. (2.9).

Dividing Eq. (4.19) by (4.18), we have

$$1 = \frac{\dot{T}_c}{\frac{\dot{T}}{\delta\sqrt{\theta}}} \tag{4.20}$$

This provides the correction we seek, i.e.

$$\dot{T}_c = \frac{\dot{T}}{\delta\sqrt{\theta}} \tag{4.21}$$

4.4 Correction of Pressure Rate of Change

Consider, as we did in Chap. 3, the pressure ratio across the Fan (inner diameter) and the Low Pressure Compressor (LPC) from station 2 to station 2.5 of our sample engine, i.e.

$$\frac{P_{25}}{P_2} = \left(\frac{T_{25}}{T_2}\right)^{\frac{\gamma\eta}{\gamma-1}} \Rightarrow \ln\left(P_{25}\right) - \ln\left(P_2\right) = \frac{\gamma\eta}{\lambda - 1}\left[\ln\left(T_{25}\right) - \ln\left(T_2\right)\right] \tag{4.22}$$

Differentiating we obtain

$$\begin{aligned} \frac{\dot{P}_{25}}{P_{25}} &= \frac{\gamma\eta}{\gamma - 1}\frac{\dot{T}_{25}}{T_{25}} \quad \text{since}\, \dot{P}_2 = 0 = \dot{T}_2 \;\; \text{implying that} \\ \dot{P}_{25} &= \frac{\gamma\eta}{\gamma - 1}\left(\frac{\dot{T}_{25}P_{25}}{T_{25}}\right) \end{aligned} \tag{4.23}$$

Likewise, this relationship (Eq. 4.23) must hold for corrected parameters as well (axiom 3),

$$\dot{P}_{25c} = \frac{\gamma\eta}{\gamma - 1}\left(\frac{\dot{T}_{25c}P_{25c}}{T_{25c}}\right) \tag{4.24}$$

Therefore, from Eq. (4.24) we can obtain

$$\begin{aligned} \frac{\sqrt{\theta}}{\delta^2}\dot{P}_{25} &= \frac{\gamma\ \eta}{\gamma - 1}\frac{\sqrt{\theta}}{\delta^2}\left(\frac{\dot{T}_{25}P_{25}}{T_{25}}\right) = \frac{\gamma\eta}{\gamma - 1}\left(\frac{\dot{T}_{25}}{\delta\sqrt{\theta}}\frac{P_{25}}{\delta}\frac{\theta}{T_{25}}\right) \\ &= \frac{\gamma\eta}{\gamma - 1}\left(\frac{\dot{T}_{25c}P_{25c}}{T_{25c}}\right) \end{aligned} \tag{4.25}$$

This gives us the required correction at station 2.5, i.e.,

$$\dot{P}_{25c} = \frac{\dot{P}_{25}\sqrt{\theta}}{\delta^2} \tag{4.26}$$

Moving further downstream in the engine and proceeding in the same manner, we have

$$\begin{aligned}
&\frac{P_3}{P_{25}} = \left(\frac{T_3}{T_{25}}\right)^{\frac{\gamma}{\gamma-1}\eta} \Rightarrow \ln(P_3) - \ln(P_{25}) = \frac{\gamma\eta}{\gamma-1}[\ln(T_3) - \ln(T_{25})] \Rightarrow \\
&\frac{\dot{P}_3}{P_3} - \frac{\dot{P}_{25}}{P_{25}} = \frac{\gamma\eta}{\gamma-1}\left(\frac{\dot{T}_3}{T_3} - \frac{\dot{T}_{25}}{T_{25}}\right) \Rightarrow \frac{\dot{P}_{3c}}{P_{3c}} - \frac{\dot{P}_{25c}}{P_{25c}} = \frac{\gamma\eta}{\gamma-1}\left(\frac{\dot{T}_{3c}}{T_{3c}} - \frac{\dot{T}_{25c}}{T_{25c}}\right) \Rightarrow \\
&\dot{P}_{3c} = P_{3c}\left[\frac{\gamma\eta}{\gamma-1}\left(\frac{\dot{T}_{3c}}{T_{3c}} - \frac{\dot{T}_{25c}}{T_{25c}}\right) + \frac{\dot{P}_{25c}}{P_{25c}}\right]
\end{aligned} \tag{4.27}$$

Multiplying by $\sqrt{\theta}/\delta$ yields

$$\begin{aligned}
&\frac{\sqrt{\theta}}{\delta}\left(\frac{\dot{P}_3}{P_3} - \frac{\dot{P}_{25}}{P_{25}}\right) = \frac{\sqrt{\theta}}{\delta}\left[\frac{\gamma\eta}{\gamma-1}\left(\frac{\dot{T}_3}{T_3} - \frac{\dot{T}_{25}}{T_{25}}\right)\right] \Rightarrow \\
&\frac{\frac{\dot{P}_3\sqrt{\theta}}{\delta^2}}{\frac{P_3}{\delta}} - \frac{\frac{\dot{P}_{25}\sqrt{\theta}}{\delta^2}}{\frac{P_{25}}{\delta}} = \frac{\gamma\eta}{\gamma-1}\left(\frac{\frac{\dot{T}_3}{\delta\sqrt{\theta}}}{\frac{T_3}{\theta}} - \frac{\frac{\dot{T}_{25}}{\delta\sqrt{\theta}}}{\frac{T_{25}}{\theta}}\right) \Rightarrow \\
&\frac{\dot{P}_3\sqrt{\theta}}{\delta^2} = P_{3c}\left[\frac{\gamma\eta}{\gamma-1}\left(\frac{\dot{T}_{3c}}{T_{3c}} - \frac{\dot{T}_{25c}}{T_{25c}}\right) + \frac{\dot{P}_{25c}}{P_{25c}}\right] = \dot{P}_{3c}
\end{aligned} \tag{4.28}$$

Thus, the same relationship holds for station 3, i.e.

$$\dot{P}_{3c} = \frac{\dot{P}_3\sqrt{\theta}}{\delta^2} \tag{4.29}$$

Repeating the same argument as we move downstream, we can conclude in general that

$$\dot{P}_c = \frac{\dot{P}\sqrt{\theta}}{\delta^2} \tag{4.30}$$

4.5 Correction of Air Flow Rate of Change

Starting with the definition of horsepower as being proportional to airflow (w_a) times the change in enthalpy (Δh), i.e.

$$
\begin{aligned}
&HP \propto (w_a \Delta h) \Rightarrow HP = const(w_a \Delta h) \\
\therefore\ &HP = const(w_a\, cp\, \Delta T) \Rightarrow \\
&HP_c = const\left(w_{ac}\, c_p\, \Delta T_c\right) \Rightarrow \\
&\dot{HP}_c = const\, \dot{w}_{ac}\, c_p\, \Delta T_c + const\, w_{ac}\, c_p\, \Delta \dot{T}_c \\
&and \\
&\frac{\dot{HP}}{\delta^2} = const\ \frac{\dot{w}_a c_p \Delta T}{\delta^2} + const \frac{w_a c_p \Delta \dot{T}}{\delta^2}
\end{aligned}
\tag{4.31}
$$

Since we have already demonstrated that $\dot{HP}_c = \dfrac{\dot{HP}}{\delta^2}$ and if $\Delta T = T - T_2$ then $\Delta \dot{T} = \dot{T} - \dot{T}_2 = \dot{T}$ and $\Delta \dot{T}_c = \dot{T}_c - \dot{T}_2 = \dot{T}_c$, we obtain

$$
const\, \dot{w}_{a\,c}\, c_p \Delta T_c + const\, w_{a\,c}\, c_p\, \Delta \dot{T}_c = const \frac{\dot{w}_a c_p \Delta T}{\delta^2} + const \frac{w_a c_p \Delta \dot{T}}{\delta^2} \tag{4.32}
$$

Cancelling like constants yields

$$
\begin{aligned}
&\dot{w}_{a\,c} \Delta T_c + w_{a\,c} \Delta \dot{T}_c = \frac{\dot{w}_a \Delta T}{\delta^2} + \frac{w_a \Delta \dot{T}}{\delta^2} \Rightarrow \\
&\dot{w}_{a\,c}\left(\frac{T}{\theta} - \frac{T_2}{\theta}\right) + \frac{w_a \sqrt{\theta}}{\delta} \dot{T}_c = \frac{\dot{w}_a}{\delta^2}(T - T_2) + \frac{w_a \dot{T}}{\delta^2}
\end{aligned}
\tag{4.33}
$$

Thus

$$
\begin{aligned}
&\dot{w}_{a\,c}\left(\frac{T}{\theta} - \frac{T_2}{\theta}\right) + \frac{w_a \sqrt{\theta}}{\delta} \frac{\dot{T}}{\delta\sqrt{\theta}} = \frac{\dot{w}_a \theta}{\delta^2}\left(\frac{T}{\theta} - \frac{T_2}{\theta}\right) + \frac{w_a \sqrt{\theta}}{\delta} \frac{\dot{T}}{\delta \sqrt{\theta}} \quad \Rightarrow \\
&\left(\dot{w}_{a\,c} - \frac{\dot{w}_a \theta}{\delta^2}\right)\left(\frac{T}{\theta} - \frac{T_2}{\theta}\right) = 0 \quad \Rightarrow \\
&\left(\dot{w}_{a\,c} - \frac{\dot{w}_a \theta}{\delta^2}\right) = 0
\end{aligned}
\tag{4.34}
$$

providing the required correction

$$
\dot{w}_{ac} = \frac{\dot{w}_a \theta}{\delta^2} \tag{4.35}
$$

4.6 Correction of Fuel Flow Rate of Change

From basic energy balance, we have the following relationship

$$
\begin{aligned}
w_f \eta_b LHV &\approx w_g(\Delta h) \approx \left(w_a + w_f\right) cp \Delta T = w_a(1 + FAR) cp \Delta T \\
&\approx w_a cp \Delta T
\end{aligned}
\tag{4.36}
$$

where η_b is burner efficiency, LHV is fuel lower heating value and FAR is fuel air ratio. Then differentiating with respect to time, we have

$$\begin{aligned}
\dot{w}_f \eta_b LHV &= \dot{w}_a cp \Delta T + w_a cp \Delta \dot{T} = cp\left(\dot{w}_a \Delta T + w_a \Delta \dot{T}\right) \Rightarrow \\
\frac{\dot{w}_f}{\delta^2} \eta_b LHV &= cp\left(\frac{\dot{w}_a \theta}{\delta^2} \frac{\Delta T}{\theta} + \frac{w_a \sqrt{\theta}}{\delta} \frac{\Delta \dot{T}}{\delta \sqrt{\theta}}\right) \\
&= cp\left(\dot{w}_{ac} \Delta T_c + w_{ac} \Delta \dot{T}_c\right) = \dot{w}_{fc} \eta_b LHV
\end{aligned} \tag{4.37}$$

Cancelling constant values we obtain our correction, i.e.,

$$\dot{w}_{fc} = \frac{\dot{w}_f}{\delta^2} \tag{4.38}$$

4.7 Correction of Net Thrust Rate of Change

In general, we can assume

$$F_n = w_g \sqrt{const\ cp\ \Delta T} - w_T \sqrt{const\ \gamma\ R}\ Mn \sqrt{T_a} \tag{4.39}$$

where

$$\begin{aligned}
T_a &= ambient\ temperature \\
w_T &= total\ airflow \\
w_{AD} &= bypass\ airflow \\
w_g &= w_T - w_{AD} + w_f \\
\Delta T &= T_s - T_a
\end{aligned}$$

Therefore

$$\begin{aligned}
\dot{F}_n &= \sqrt{const\, cp}\left[\dot{w}_g \sqrt{\Delta T} + \frac{1}{2} w_g \frac{\Delta \dot{T}}{\sqrt{\Delta T}}\right] - \sqrt{const \gamma R Mn}\left[\dot{w}_T \sqrt{T_a} + \frac{1}{2} w_T \frac{\dot{T}_a}{\sqrt{T_a}}\right] \\
&= const\left[\dot{w}_g \sqrt{\Delta T} + \frac{1}{2} w_g \frac{\Delta \dot{T}}{\sqrt{\Delta T}}\right] - const\left[\dot{w}_T \sqrt{T_a} + \frac{1}{2} w_T \frac{\dot{T}_a}{\sqrt{T_a}}\right] \Rightarrow \\
\frac{\dot{F}_n \sqrt{\theta}}{\delta^2} &= const\left[\frac{\dot{w}_g \theta}{\delta^2} \sqrt{\frac{\Delta T}{\theta}} + \frac{1}{2} \frac{w_g \sqrt{\theta}}{\delta} \frac{\Delta \dot{T}}{\sqrt{\theta} \delta} \sqrt{\frac{\theta}{\Delta T}}\right] \\
&\quad - const\left[\frac{\dot{w}_T \theta}{\delta^2} \sqrt{\frac{T_a}{\theta}} + \frac{1}{2} \frac{w_T \sqrt{\theta}}{\delta} \frac{\dot{T}_a}{\sqrt{\theta} \delta} \sqrt{\frac{\theta}{T_a}}\right] \\
&= const\left[\dot{w}_{gc} \sqrt{\Delta T_c} + \frac{1}{2} w_{gc} \frac{\Delta \dot{T}_c}{\sqrt{\Delta T_c}}\right] - const\left[\dot{w}_{Tc} \sqrt{T_{ac}} + \frac{1}{2} w_{Tc} \frac{\dot{T}_{ac}}{\sqrt{T_{ac}}}\right]
\end{aligned}$$

Everything on the right hand side of the last equation is in terms of corrected quantities and hence constant at an arbitrary operating point and so

$$\frac{\dot{F}_n\sqrt{\theta}}{\delta^2} = const \tag{4.40}$$

which provides our correction, i.e.

$$\dot{F}_{nc} = \frac{\dot{F}_n\sqrt{\theta}}{\delta^2} \tag{4.41}$$

4.8 Correction of Fuel Air Ratio (FAR) Rate of Change

By definition, we have

$$FAR = \frac{w_f}{w_a} \tag{4.42}$$

By axiom 3, we must have the same relationship for corrected quantities, i.e.

$$FAR_c = \frac{w_{fc}}{w_{ac}} = \frac{\dfrac{w_f}{\delta\sqrt{\theta}}}{\dfrac{w_a\sqrt{\theta}}{\delta}} = \frac{\dfrac{w_f}{w_a}}{\theta} = \frac{FAR}{\theta} \tag{4.43}$$

Equation (4.16) provides the correction for FAR. If we differentiate Eq. (4.42), we obtain

$$F\dot{A}R = \frac{\dot{w}_f w_a - w_f \dot{w}_a}{w_a^2} \tag{4.44}$$

Dividing Eq. (4.44) by $\delta/\sqrt{\theta}$ yields

$$\frac{F\dot{A}R}{\delta\sqrt{\theta}} = \frac{\dfrac{\dot{w}_f}{\delta^2}\dfrac{w_a\sqrt{\theta}}{\delta} - \dfrac{w_f}{\delta\sqrt{\theta}}\dfrac{\dot{w}_a\theta}{\delta^2}}{\dfrac{w_a^2\theta}{\delta^2}} = \frac{\dot{w}_{fc}w_{ac} - w_{fc}\dot{w}_{ac}}{w_{a\,c}^2} = F\dot{A}R_c \tag{4.45}$$

Proving that

$$\dot{FAR}_c = \frac{\dot{FAR}}{\delta\sqrt{\theta}} \tag{4.46}$$

4.9 Correction of Thrust Specific Fuel Consumption (TSFC) Rate of Change

Likewise we know

$$TSFC = \frac{w_f}{F_n} \Rightarrow \frac{TSFC}{\sqrt{\theta}} = \frac{\frac{w_f}{\delta\sqrt{\theta}}}{\frac{F_n}{\delta}} = TSFC_c \tag{4.47}$$

Differentiating Eq. (4.47)

$$\begin{aligned} TS\dot{F}C &= \frac{\dot{w}_f F_n - w_f \dot{F}_n}{{F_n}^2} \qquad \Rightarrow \\ \frac{TS\dot{F}C}{\delta} &= \frac{\frac{\dot{w}_f}{\delta^2}\frac{F_n}{\delta} - \frac{w_f}{\delta\sqrt{\theta}}\frac{\dot{F}_n\sqrt{\theta}}{\delta^2}}{\frac{{F_n}^2}{\delta^2}} = \frac{\dot{w}_{fc} F_{nc} = w_{fc}\dot{F}_{nc}}{F_{nc}^2} = TS\dot{F}C_c \end{aligned} \tag{4.48}$$

yielding the required correction

$$TS\dot{F}C_c = \frac{TS\dot{F}C}{\delta} \tag{4.49}$$

4.10 Summary of Rate of Change Corrections

As a result of these derivations, there is an interesting observation that can be made. We list below the time derivative corrections derived in this Chapter:

$$\dot{SHP}_c = \left(\frac{\dot{SHP}}{\delta^2}\right) = \frac{\sqrt{\theta}}{\delta}\frac{d}{dt}\left(\frac{SHP}{\delta\sqrt{\theta}}\right) = \left[\frac{\sqrt{\theta}}{\delta}\right]\frac{d}{dt}(SHP_c)$$

$$\dot{Q}_c = \left(\frac{\dot{Q}\sqrt{\theta}}{\delta^2}\right) = \frac{\sqrt{\theta}}{\delta}\frac{d}{dt}\left(\frac{Q}{\delta}\right) = \left[\frac{\sqrt{\theta}}{\delta}\right]\frac{d}{dt}(Q_c)$$

$$\dot{w}_{a\,c} = \left(\frac{\dot{w}_a\theta}{\delta^2}\right) = \frac{\sqrt{\theta}}{\delta}\frac{d}{dt}\left(\frac{w_a\sqrt{\theta}}{\delta}\right) = \left[\frac{\sqrt{\theta}}{\delta}\right]\frac{d}{dt}(w_{a\,c})$$

$$\dot{T}_c = \left(\frac{\dot{T}}{\delta\sqrt{\theta}}\right) = \frac{\sqrt{\theta}}{\delta}\frac{d}{dt}\left(\frac{T}{\theta}\right) = \left[\frac{\sqrt{\theta}}{\delta}\right]\frac{d}{dt}(T_c)$$

$$\dot{P}_c = \left(\frac{\dot{P}\sqrt{\theta}}{\delta^2}\right) = \frac{\sqrt{\theta}}{\delta}\frac{d}{dt}\left(\frac{P}{\delta}\right) = \left[\frac{\sqrt{\theta}}{\delta}\right]\frac{d}{dt}(P_c)$$

$$\dot{N}_c = \left(\frac{\dot{N}}{\delta}\right) = \frac{\sqrt{\theta}}{\delta}\frac{d}{dt}\left(\frac{N}{\sqrt{\theta}}\right) = \left[\frac{\sqrt{\theta}}{\delta}\right]\frac{d}{dt}(N_c)$$

$$\ddot{N}_c = \left(\frac{\ddot{N}\sqrt{\theta}}{\delta^2}\right) = \frac{\sqrt{\theta}}{\delta}\frac{d}{dt}\left(\frac{\dot{N}}{\delta}\right) = \left[\frac{\sqrt{\theta}}{\delta}\right]\frac{d}{dt}(\dot{N}_c)$$

$$\dot{w}_{f\,c} = \frac{\dot{w}_f}{\delta^2} = \frac{\sqrt{\theta}}{\delta}\frac{d}{dt}\left(\frac{w_f}{\delta\sqrt{\theta}}\right) = \left[\frac{\sqrt{\theta}}{\delta}\right]\frac{d}{dt}(w_{f\,c})$$

$$\dot{F}_{n\,c} = \frac{\dot{F}_n\sqrt{\theta}}{\delta^2} = \frac{\sqrt{\theta}}{\delta}\frac{d}{dt}\left(\frac{F_n}{\delta}\right) = \left[\frac{\sqrt{\theta}}{\delta}\right]\frac{d}{dt}(F_{n\,c})$$

$$\dot{TSFC}_c = \frac{\dot{TSFC}}{\delta} = \frac{\sqrt{\theta}}{\delta}\frac{d}{dt}\left(\frac{TSFC}{\sqrt{\theta}}\right) = \left[\frac{\sqrt{\theta}}{\delta}\right]\frac{d}{dt}(TSFC_c)$$

$$\dot{FAR}_c = \frac{\dot{FAR}}{\delta\sqrt{\theta}} = \frac{\sqrt{\theta}}{\delta}\frac{d}{dt}\left(\frac{FAR}{\theta}\right) = \left[\frac{\sqrt{\theta}}{\delta}\right]\frac{d}{dt}(FAR_c)$$

From the list above, a simple relationship seems to emerge, that for a gas path parameter X, we have

$$\dot{X}_c = \frac{\sqrt{\theta}}{\delta}\frac{d}{dt}(X_c) \tag{4.50}$$

We have established that Eq. (4.50) holds for temperatures, pressures, speeds, flows, power, i.e., all common gas path parameters of interest. We have not, however, formally demonstrated this relationship, for an *arbitrary* parameter X, as resulting from the axioms of Chap. 2. As such, we will use Eq. (4.50) as a *conjectured* result except for the parameters noted in this Chapter. For the latter, this relationship provides a simple and quick method for obtaining the form of the time derivative correction given the knowledge of the fundamental parameter correction. We will also make use of this conjecture in the next Chapter as we explore the correction of other ancillary parameters of interest.

Chapter 5
Additional Corrections

Abstract This chapter provides the Standard Day corrections for several additional parameters not already considered in previous chapters. These include Time Constants, Metal Temperature, and Static Temperature. The derivations proceed as before, employing only the definitions, axioms, assumptions, and derived results from preceding chapters. It should be noted that an additional assumption is employed in the derivation for Metal Temperature, i.e., the *conjectured* result given in Eq. (4.50).

Keywords Corrections · Time constants · Metal temperature · Static temperature · Theta · Delta · Specific heat · Local Mach No. · Ratio of specific heats · Linear system

In this Chapter we will explore the corrections for three additional parameters, namely, (a) time constants, (b) metal temperature, and (c) static temperature. Again we will use the axioms of Chap. 2 and additionally, make use of the *conjectured* relationship given in Eq. (4.50) of the last Chapter.

5.1 Correction of Time Constants

Suppose we consider a first order linear system with time constant τ, where X is an arbitrary gas path parameter and X_{ss} represents the steady state value of X after transient. Mathematically, this is represented as follows:

$$\dot{X} = \frac{1}{\tau}(X - X_{ss}) \tag{5.1}$$

It is a fundamental assumption that this general relationship will hold (in form) at ISA conditions, therefore

A. J. Volponi, *Gas Turbine Parameter Corrections*,
https://doi.org/10.1007/978-3-030-41076-6_5

$$\dot{X}_c = \frac{1}{\tau_c}(X_c - X_{ssc}) \tag{5.2}$$

where the subscript *c* denotes corrected parameter. It is reasonable to assume that the time constant value τ_c may be different than the time constant value τ, since certainly for temperatures ($X = T$), it is known that τ will vary with (air) flow which in turn varies with ambient temperature & pressure ($\tau \downarrow$ as $w_a \uparrow$ as $\theta_2\downarrow$). From Chap. 2, we know the general form of a corrected parameter to be (Eq. 2.2)

$$X_c = \frac{X}{\theta^a \delta^b}$$

Applying the relationship from Eq. (4.50) we obtain

$$\dot{X}_c = \frac{\sqrt{\theta}}{\delta}\frac{d}{dt}(X_c) \Rightarrow \dot{X}_c = \frac{\sqrt{\theta}}{\delta}\left(\frac{\dot{X}}{\theta^a \delta^b}\right) \tag{5.3}$$

Therefore

$$\frac{\sqrt{\theta}}{\delta}\left(\frac{\dot{X}}{\theta^a \delta^b}\right) = \dot{X}_c = \frac{1}{\tau_c}\left(\frac{X}{\theta^a \delta^b} - \frac{X_{ss}}{\theta^a \delta^b}\right) \Rightarrow \frac{\sqrt{\theta}}{\delta}\dot{X} = \frac{1}{\tau_c}(X - X_{ss}) \tag{5.4}$$

A little manipulation yields

$$\frac{1}{\tau_c}(X - X_{ss}) = \frac{\sqrt{\theta}}{\delta}\frac{1}{\tau}(X - X_{ss}) \tag{5.5}$$

which provides the correction we seek

$$\tau_c = \left(\frac{\delta}{\sqrt{\theta}}\right)\tau \tag{5.6}$$

5.2 Correction of Metal Temperature

At this juncture we will derive a correction for metal temperatures using results from Chap. 3 and the *conjectured* relationship, Eq. (4.50), for an arbitrary time derivative correction. From Chap. 3, (Eq. 3.46), we know that the requisite correction for the time derivative of metal temperature T_m is as follows:

$$\dot{T}_{mc} = \frac{\dot{T}_m}{\theta^{0.74}\delta^{0.8}}$$

If we use this relationship in conjunction with Eq. (4.50), we have the following:

$$\frac{\dot{T}_m}{\theta^{0.74}\delta^{0.8}} = \dot{T}_{mc} = \frac{\sqrt{\theta}}{\delta}\frac{d}{dt}(T_{mc}) = \frac{\sqrt{\theta}}{\delta}\frac{d}{dt}\left(\frac{T_m}{\theta^a\delta^b}\right) = \frac{\dot{T}_m}{\theta^{a-0.5}\delta^{b+1}} \tag{5.7}$$

where a and b are the required correction exponents to standardize the metal temperature. Equating the left and right hand sides of this equation provides the two equations

$$0.74 = a - 0.5 \quad and \quad 0.8 = b + 1 \Rightarrow a = 1.24 \quad and \quad b = -0.2$$

This provides a correction for metal temperature as

$$T_{mc} = \frac{T_m\delta^{0.2}}{\theta^{1.24}} \tag{5.8}$$

Since gas temperature is corrected (classically) by θ, i.e. T/θ, we see that there is an additional multiplicative factor α_m as follows

$$T_{mc} = \frac{T_m\delta^{0.2}}{\theta^{1.24}} \approx \left(\frac{\delta}{\theta}\right)^{0.2}\frac{T_m}{\theta} \equiv \alpha_m\frac{T_m}{\theta} \tag{5.9}$$

The magnitude of this multiplicative factor will depend on the flight condition (altitude and Mach no.) and for ISA conditions will vary as illustrated in the following Fig. 5.1.

The corrections for Metal Temperatures and Spool Speeds have implications in gas turbine engine modeling, in particular with simple piecewise linear state representation models that are used in real-time (on-board) applications. In these types of models, the Metal Temperatures and Spool Speeds typically form the *state* parameters in the model which in its simplest formulation has the following basic form

$$\dot{x} = A\Delta x + B\Delta u$$

$$\Delta y = C\Delta x + D\Delta u$$

The term Δ indicates delta from steady state condition. The terms A, B, C, and D are matrices of partial derivatives, Δx is a vector of state parameters (combination of Metal Temperatures (T_m) and Spool Speeds, N_1 and N_2), Δu is a vector of input parameters (e.g. Fuel Flow, Bleed and variable geometry commands) and Δy is a vector of measured (and un-measured) dependent engine gas path parameters (e.g. gas temperatures, pressures, flows, etc). To minimize storage and model size, these elements are typically chosen to be standard day *corrected* parameters. To solve these equations, at each time step, integration (of the state variables) must be performed and hence the corrected parameter derivatives must be *un-corrected,* integrated, and the resultant (raw) states *re-corrected*, i.e.

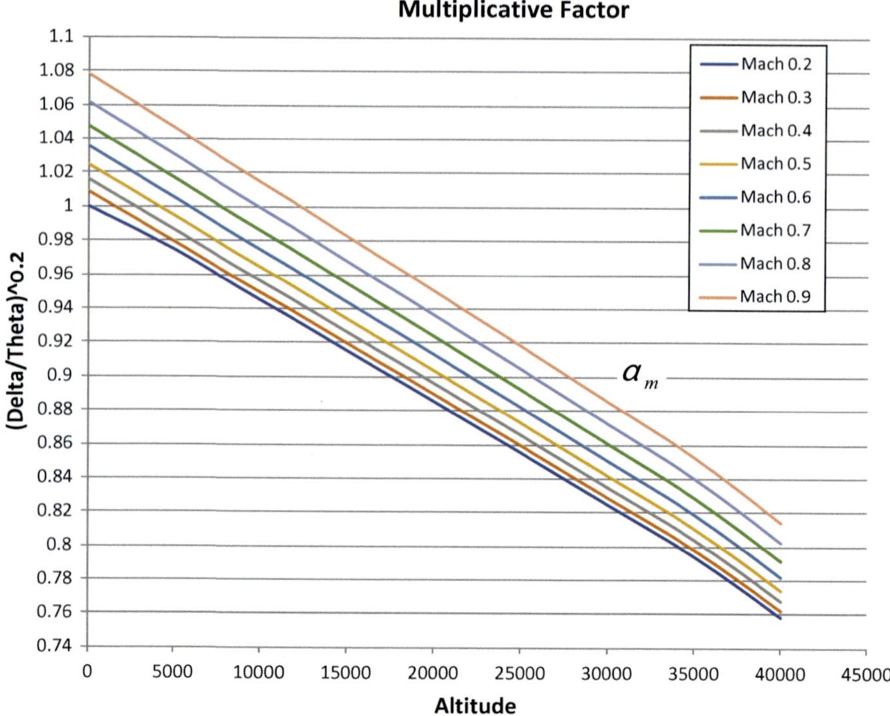

Fig. 5.1 Multiplicative factor α_m

$$\dot{N}_c \rightarrow \delta \rightarrow \dot{N} \rightarrow \int \rightarrow N \rightarrow \frac{1}{\sqrt{\theta}} \rightarrow N_c$$

$$\dot{T}_{mc} \rightarrow \theta^{0.74}\delta^{0.8} \rightarrow \dot{T}_m \rightarrow \int \rightarrow T_m \rightarrow \frac{\delta^{0.2}}{\theta^{1.24}} \rightarrow T_{mc}$$

5.3 Correction of Static Temperature

We will now explore to what degree, if any, a static temperature's correction differs from that of a total temperature. We start with the familiar relationship between static and total temperature and their relationship to local Mach no., i.e.

$$\frac{T}{T_s} = 1 + \frac{\gamma - 1}{2} M_n^2 \tag{5.10}$$

If we assume, (as we do for the *classical* corrections in Chap. 3) that the ratio of specific heats, γ, is constant and that local Mach No. is constant, then we see that

$$\frac{dT_s}{T_s} = \frac{dT}{T} \Rightarrow \frac{T_s}{\theta} = \frac{T}{\theta} = T_c = const \Rightarrow T_{sc} = \frac{T_s}{\theta} \tag{5.11}$$

Under these (classical) assumptions, the correction for static and total temperature is the same. We will now relax some of the assumptions and simplifications. Returning to Eq. (5.10), if we take logs of both sides and differentiate, we obtain

$$\begin{aligned}\frac{dT}{T} - \frac{dT_s}{T_s} &= \frac{d\left(1 + \frac{\gamma-1}{2}M_n^2\right)}{1 + \frac{\gamma-1}{2}M_n^2} = \frac{2M_n\left(\frac{\gamma-1}{2}\right)dM_n + \frac{1}{2}M_n^2 d\gamma}{1 + \frac{\gamma-1}{2}M_n^2} \\ &= \frac{M_n^2(\gamma - 1)\frac{dM_n}{M_n} + \frac{1}{2}M_n^2 d\gamma}{1 + \frac{\gamma-1}{2}M_n^2}\end{aligned} \tag{5.12}$$

Under our axiomatic assumption that local Mach No. is constant, this reduces to

$$\frac{dT}{T} - \frac{dT_s}{T_s} = \underbrace{\left[\frac{M_n^2\gamma}{2\left(1 + \frac{\gamma-1}{2}M_n^2\right)}\right]}_{\alpha}\frac{d\gamma}{\gamma} = \alpha\frac{d\gamma}{\gamma} \tag{5.13}$$

The multiplicative factor, α, will vary with the absolute level of local Mach No and the ratio of specific heats. Figure 5.2 below depicts how this factor varies.

This plot portrays the multiplicative factor as a function of local Mach no., wherein the solid line is the average factor (across varying γ) and the dashed lines

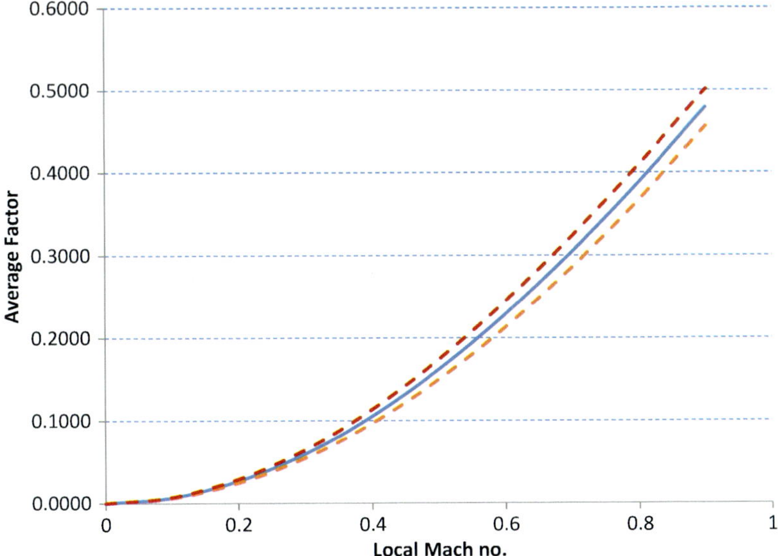

Fig. 5.2 Multiplicative factor α

indicate ±3 sigma range. For static air, i.e. Mach = 0, the factor is zero and static temperature equals static temperature as expected. For most commercial gas turbines, in areas where temperature *measurements* are typical, the local Mach no. is approximately in the range of 0.3–0.4, giving α a range of values between 0.06 and 0.10. For illustration, let us assume a local Mach No. value, in that range, of 0.35, which gives an average value of $\alpha = 0.08$. Thus

$$\frac{dT}{T} - \frac{dT_s}{T_s} = 0.08\frac{d\gamma}{\gamma} \tag{5.14}$$

In addition, we can approximate γ as a function of T_s as $\gamma \approx 1.8191(T_s)^{-0.041}$ (with $R^2 = 0.9866$), which implies

$$\frac{d\gamma}{\gamma} \approx -0.041\frac{dT_s}{T_s} \tag{5.15}$$

Therefore, we have the approximation

$$\frac{dT}{T} - \frac{dT_s}{T_s} \approx 0.08\frac{d\gamma}{\lambda} \approx -0.00328\frac{dT_s}{T_s} \Rightarrow \frac{dT}{T} \approx .99672\frac{dT_s}{T_s} \Rightarrow \frac{dT_s}{T_s} \approx 1.0033\frac{dT}{T} \tag{5.16}$$

Now if α_T represents the theta exponent for T, (classically $\alpha_T = 1$),[1] then

$$\frac{dT_s}{T_s} \approx 1.0033\frac{dT}{T} = 1.0033\alpha_T\frac{d\theta}{\theta} \tag{5.17}$$

from which it follows

$$T_{sc} \approx \frac{T_s}{\theta^{1.0033\alpha_T}} \tag{5.18}$$

The effect of relaxing our classical assumption, clearly, has very little impact for low local Mach numbers. If we consider much higher Mach numbers (around 0.8) the theta exponent increases to approximately $1.016\alpha_T$ which is a bit more significant.

In the next Chapter we will continue to derive some refinements to the classical corrections given in Chap. 3 by relaxing some of our assumptions, in particular the assumption that specific heats are constant.

[1]We will explore the use of non-unity theta exponent corrections for Temperatures in Chap. 7 on empirical methods.

Chapter 6
Refinements to Common Corrections

Abstract This chapter provides some refinement to the Standard Day corrections for most of the common gas path parameters considered in Chap. 3. In particular, the parameters considered will include, Rotor Speed, Pressure, Air Flow, Fuel Flow, Horsepower, Torque, and Acceleration. The refinements will stem from the *relaxation* of our previous assumption that specific heats, *cp* and *cv*, are constant and do not change with temperature. Allowing specific heats (and their ratio $\gamma = cp/cv$) to vary, as they naturally do, offers theta exponents that are slightly different than the classical values provided in Chap. 3. The derivations follow the same strategy as in Chap. 3 and abide by the constraints imposed by our definitions, axioms, and previous results.

Keywords Refinements · Corrections · Rotor speed · Pressure · Air flow · Fuel flow · Horsepower · Torque · Acceleration · Theta · Delta · Specific heat variation · Local Mach No. · Ratio of specific heats

This chapter explores some refinements to the classical corrections given in Chap. 3. This may provide some additional accuracy in applications where corrected parameters are being employed. The level of improvement would depend on a number of factors including the particular application at hand, thus it is not possible to estimate the actual level of improvement a-priori. The chapter, however, will provide the correction enhancements and the practitioner can evaluate their impact on an application by application basis.

Improvement may be possible if *we relax some of our assumptions, in particular, the assumption of constant specific heats, cp and cv.* To explore this possibility, we begin by re-considering the correction for rotational spool speed as a first example.

A. J. Volponi, *Gas Turbine Parameter Corrections*,
https://doi.org/10.1007/978-3-030-41076-6_6

6.1 Speed Correction

Tangential velocity is related to rotational speed (rpm) by the radius of the object (compressor or turbine blade) in question. It is also related to acoustic velocity by Mach number and the square root of temperature. In symbols

$$v = rN = M_n\sqrt{\gamma R T_s} \tag{6.1}$$

where v is the velocity in m/s, r is the radius (m), N is the rotational speed (rad/s), γ, and R are the ratio of specific heats and the gas constant respectively, and M_n is local Mach number. Taking logs and differentiating, we obtain from Eq. (6.1)

$$\frac{dr}{r}+\frac{dN}{N}=\frac{dM_n}{M_n}+\frac{1}{2}\left(\frac{d\gamma}{\gamma}+\frac{dR}{R}+\frac{dT_s}{T_s}\right)\Rightarrow\frac{dN}{N}=\frac{1}{2}\left(\frac{d\gamma}{\gamma}+\frac{dT_s}{T_s}\right) \tag{6.2}$$

But since

$$\frac{T}{T_s}=1+\frac{\gamma-1}{2}M_n^2\approx const\Rightarrow\frac{dT}{T}\approx\frac{dT_s}{T_s} \tag{6.3}$$

Hence

$$\frac{dN}{N}\approx\frac{1}{2}\left(\frac{d\gamma}{\gamma}+\frac{dT}{T}\right) \tag{6.4}$$

Since $\gamma = cp/cv$ we have that

$$\frac{d\gamma}{\gamma}=\frac{dcp}{cp}-\frac{dcv}{cv} \tag{6.5}$$

We also recall $R = cp - cv$ and therefore

$$\begin{gathered}0=\frac{dR}{R}=\frac{dcp}{cp-cv}-\frac{dcv}{cp-cv}=\left(\frac{cp}{cp-cv}\right)\frac{dcp}{cp}-\left(\frac{cv}{cp-cv}\right)\frac{dcv}{cv}\Rightarrow\\ \frac{dcp}{cp}=\left(\frac{cv}{cp}\right)\frac{dcv}{cv}=\frac{1}{\gamma}\frac{dcv}{cv}<\frac{dcv}{cv}\quad\text{since } \gamma>\end{gathered} \tag{6.6}$$

Combining with Eq. (6.5) we obtain

$$\frac{d\gamma}{\gamma}=\frac{dc_p}{c_p}-\frac{dc_v}{c_v}<0 \tag{6.7}$$

If we expand Eq. (6.2) and use this last inequality we see that

$$\frac{dN}{N} \approx \frac{1}{2}\left(\frac{d\gamma}{\gamma} + \frac{dT}{T}\right) = \frac{1}{2}\left[\frac{\frac{d\gamma}{\gamma} + \frac{dT}{T}}{\frac{dT}{T}}\frac{dT}{T}\right] \equiv \frac{\alpha}{2}\frac{dT}{T} \quad \text{where } \alpha < 1 \quad \text{therefore}$$

$$\frac{N}{T^{\alpha/2}} = \text{constant} \Rightarrow \frac{N}{\theta^{\alpha/2}} = \text{constant}, \quad \text{with } \alpha/2 < 1/2 \tag{6.8}$$

This simple analysis indicates that if we take into account the changes in specific heat (and hence the changes in the ratio of specific heats) that the familiar classical correction of rotor speed should be replaced by an exponent less than ½, i.e. less than square root. To determine how much less, we need to approximate how the ratio of specific heats change (with temperature).

We will approximate the ratio of specific heat, γ, as follows

$$\gamma \approx \alpha T_s^{\beta} \Rightarrow \ln(\gamma) = \ln(\alpha) + \beta \ln(T_s) \Rightarrow \frac{d\gamma}{\gamma} = \beta\frac{dT_s}{T_s} \tag{6.9}$$

Figure 6.1 below, depicts this relationship along with an exponential curve fit. Hence $\beta = -0.041$ and we have

$$\begin{aligned} \frac{dN}{N} &\approx \frac{1}{2}\left(\frac{d\gamma}{\gamma} + \frac{dT_s}{T_s}\right) \\ &\approx \frac{1}{2}\left(-0.041\frac{dT_s}{T_s} + \frac{dT_s}{T_s}\right) = \frac{0.959}{2}\frac{dT_s}{T_s} \\ &\approx 0.4795\frac{dT_s}{T_s} \approx (.4795)(1.0033)\alpha_T\frac{d\theta}{\theta} \end{aligned} \tag{6.10}$$

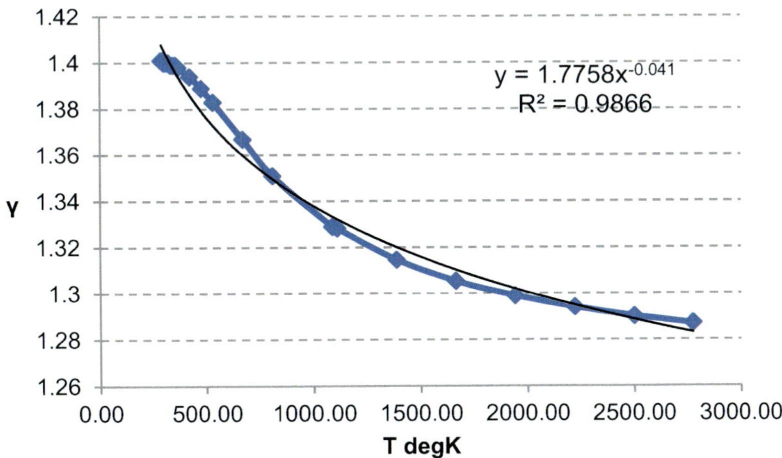

Fig. 6.1 Ratio of specific heats (γ) as a function of temperature

where the last term in this expression utilizes Eq. (5.18) from the preceding Chapter and α_T is the correction theta exponent for temperature (classically $\alpha_T = 1$). We then arrive at the refined correction for rotor speed as

$$N_c = \frac{N}{\theta^{.48\alpha_T}} = \frac{N}{\theta^{.48}} \tag{6.11}$$

6.2 Pressure Correction

We begin as we did in Chap. 3 by considering, for specificity, the pressure rise across the Low Pressure Compressor (LPC), and its relationship to temperature ratio, namely

$$\frac{T_{2.5}}{T_2} = \left(\frac{P_{2.5}}{P_2}\right)^{\frac{\gamma-1}{\gamma\eta_{25}}} \quad \text{where } \eta_{25} = \text{polytropic efficiency accross LPC}$$
$$\gamma = cp/cv = \text{ratio of specific heats} \tag{6.12}$$

Taking logs of both sides of Eq. (6.12) and rearranging terms, we have

$$\ln\left(\frac{P_{2.5}}{P_2}\right) = \frac{\gamma\eta_{25}}{\gamma-1}\ln\left(\frac{T_{2.5}}{T_2}\right) \tag{6.13}$$

Differentiating this last term provides

$$\frac{dP_{2.5}}{P_{2.5}} - \frac{dP_2}{P_2} = \frac{\gamma\eta_{25}}{\gamma-1}\left(\frac{dT_{2.5}}{T_{2.5}} - \frac{dT_2}{T_2}\right) + d\left[\frac{\gamma\eta_{25}}{\gamma-1}\right]\ln\left(\frac{T_{2.5}}{T_2}\right) \tag{6.14}$$

Recalling from elementary calculus that $d\left(\frac{u}{v}\right) = \frac{vdu - udv}{v^2}$, we can evaluate the expression

$$d\left[\frac{\gamma\eta_{25}}{\gamma-1}\right] = \frac{(\gamma-1)\eta_{25}d\gamma - \gamma\eta_{25}d\gamma}{(\gamma-1)^2} = \frac{-\eta_{25}d\gamma}{(\gamma-1)^2} = \left(\frac{-\gamma\eta_{25}}{(\gamma-1)^2}\right)\frac{d\gamma}{\gamma} \tag{6.15}$$

From the previous Chapter, we can assume that $T_{2.5}/\theta^{\alpha_{T25}} = const$, which implies that

$$dT_{2.5}/T_{2.5} = \alpha_{T25}\left(dT_2/T_2\right) \tag{6.16}$$

Combining this last equation with Eqs. (6.14) and (6.15) provides

$$\begin{aligned}\frac{dP_{2.5}}{P_{2.5}}-\frac{dP_2}{P_2}&=\frac{\gamma\eta_{25}}{\gamma-1}\left(\frac{dT_{2.5}}{T_{2.5}}-\frac{dT_2}{T_2}\right)+d\left[\frac{\gamma\eta_{25}}{\gamma-1}\right]\ln\left(\frac{T_{2.5}}{T_2}\right)\\&=\underbrace{\frac{\gamma\eta_{25}}{\gamma-1}(\alpha_{T25}-1)}_{\alpha}\frac{dT_2}{T_2}-\underbrace{\left(\frac{\gamma\eta_{25}}{(\gamma-1)^2}\right)\ln\left(\frac{T_{2.5}}{T_2}\right)}_{\beta}\frac{d\gamma}{\gamma}\\&=\alpha\frac{dT_2}{T_2}-\beta\frac{d\gamma}{\gamma}\end{aligned} \tag{6.17}$$

With a little more manipulation, we form the basis of a theta correction for pressure, i.e.

$$\begin{aligned}\frac{dP_{2.5}}{P_{2.5}}-\frac{dP_2}{P_2}&=\alpha\frac{dT_2}{T_2}-\frac{\beta\frac{d\gamma}{\gamma}}{\frac{dT_2}{T_2}}\frac{dT_2}{T_2}\\&=\underbrace{\left(\alpha-\frac{\beta\frac{d\gamma}{\gamma}}{\frac{dT_2}{T_2}}\right)}_{\alpha_{P25}}\frac{dT_2}{T_2}=\alpha_P\frac{d\theta}{\theta}\end{aligned} \tag{6.18}$$

This last expression gives the relationship

$$\frac{dP_{2.5}}{P_{2.5}}-\frac{dP_2}{P_2}=\alpha_{P25}\frac{d\theta}{\theta}\Rightarrow$$

$$P_{25c}=\frac{P_{2.5}}{\theta^{\alpha_{p25}}\delta}\quad with\quad \alpha_{P25}\approx\frac{\gamma\eta_{25}}{\gamma-1}(\alpha_{T25}-1)-\left[\left(\frac{\gamma\eta_{25}}{(\gamma-1)^2}\right)\ln\left(\frac{T_{2.5}}{T_2}\right)\right]\left(\frac{d\gamma/\gamma}{dT_2/T_2}\right) \tag{6.19}$$

If we look at how the ratio of specific heat varies with temperature (see Fig. 6.2 below where we plot the log of γ versus the log of temperature), we can state approximately that the last term

$$\left(\frac{d\gamma/\gamma}{dT_2/T_2}\right)=\frac{d\ln(\gamma)}{d\ln(T_2)}\approx\frac{\Delta\ln(\gamma)}{\Delta\ln(T_2)}\approx-0.041 \tag{6.20}$$

This provides the final general expression

$$P_{2.5c}\approx\frac{P_{2.5}}{\delta\theta^{\frac{\gamma\eta_{25}}{\gamma-1}(\alpha_{T25}-1)+.041\left[\left(\frac{\gamma\eta_{25}}{(\gamma-1)^2}\right)\ln\left(\frac{T_{2.5}}{T_2}\right)\right]}} \tag{6.21}$$

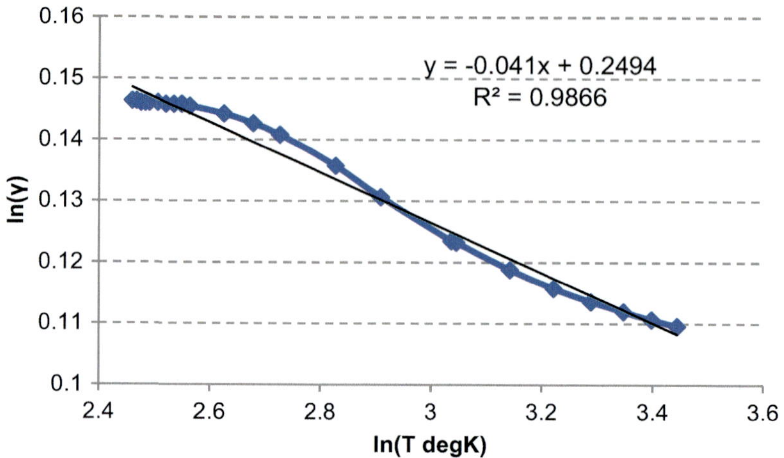

Fig. 6.2 Natural log of γ versus log of temperature

For example, if we let

$$\gamma = 1.4, \quad \eta_{25} = 0.87, \quad \alpha_{T25} = 0.93 \quad T_{2.5}/T_2 \approx 2.6 \Rightarrow \alpha_{P25} = -0.093 \Rightarrow P_{2.5c} = \frac{P_{2.5}\theta^{.093}}{\delta}$$

If we use the classical temperature theta exponent of 1, we obtain a different correction, i.e.,

$$\gamma = 1.4, \quad \eta_{25} = 0.87, \quad \alpha_{T25} = 1.0 \quad T_{2.5}/T_2 \approx 2.6 \Rightarrow \alpha_{P25} = 0.131 \Rightarrow P_{2.5c} = \frac{P_{2.5}}{\delta\theta^{0.131}}$$

Following the same arguments across the HPC (station 3) we would obtain a similar form to Eq. (6.21), namely

$$P_{3c} \approx \frac{P_3}{\delta\theta^{\frac{\gamma\eta_3}{\gamma-1}(\alpha_{T3}-1)+.041\left[\left(\frac{\gamma\eta_3}{(\gamma-1)^2}\right)\ln\left(\frac{T_3}{T_{2.5}}\right)\right]}} \tag{6.22}$$

Thus, in general, we have the correction

$$P_c \approx \frac{P}{\delta\theta^{\frac{\gamma\eta}{\gamma-1}(\alpha_T-1)+.041\left[\left(\frac{\gamma\eta}{(\gamma-1)^2}\right)\ln\left(\frac{T^*}{T}\right)\right]}} \tag{6.23}$$

where T^*/T represents the temperature ratio across the compressor or turbine in question. The range of exponents depends on the factors involved, (efficiency level,

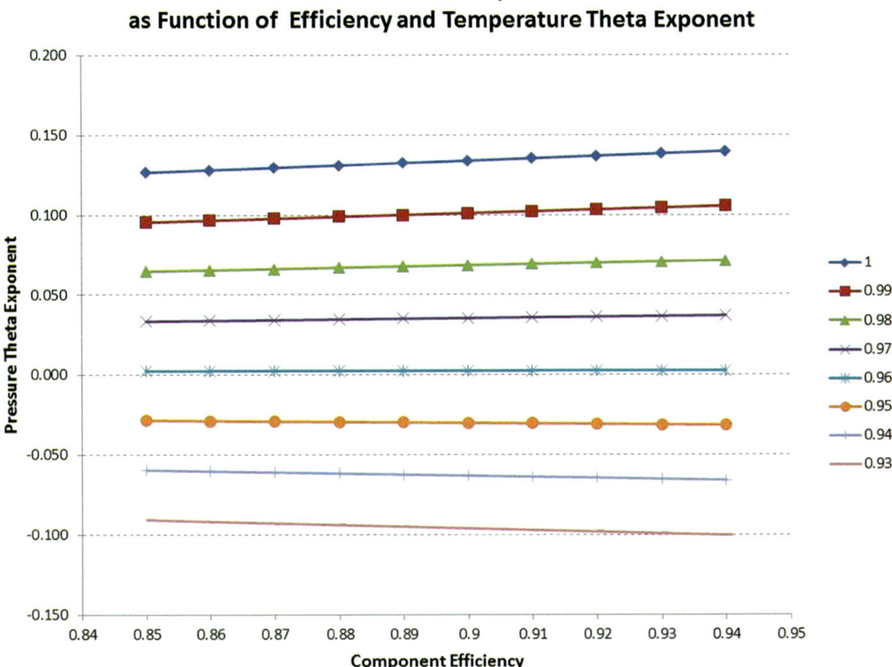

Fig. 6.3 Sample pressure theta exponents

temperature ratio and temperature theta exponent). A sample is depicted in Fig. 6.3 above for a temperature ratio of 2.6.

It is clear from the plots that component efficiency has little effect on the pressure theta exponent.

6.3 Airflow Correction

For an ideal gas, we have

$$\frac{w_a}{A} = \rho v = \frac{P_s}{RT_s} v \quad where \quad M_n = v/\sqrt{\gamma RT_s} \tag{6.24}$$

Re-writing

$$\frac{w_a}{A} = \frac{P_s M_n \sqrt{\gamma RT_s}}{RT_s} \Rightarrow \frac{w_a \sqrt{T_t}}{AP_t} = \left(\frac{P_s}{P_t}\right) \sqrt{\frac{T_t}{T_s}} M_n \sqrt{\frac{\gamma}{R}} \tag{6.25}$$

We know, however, that for an isentropic process

$$\frac{T_t}{T_s} = 1 + \frac{\gamma - 1}{2} M_n{}^2 \quad \text{and} \quad \frac{P_t}{P_s} = \left(1 + \frac{\gamma - 1}{2} M_n{}^2\right)^{\frac{\gamma}{\gamma - 1}} \tag{6.26}$$

Substituting Eq. (6.26) into Eq. (6.25), we obtain

$$\frac{w_a\sqrt{T_t}}{AP_t} = \sqrt{\frac{\gamma}{R}} M_n \left(1 + \frac{\gamma - 1}{2} M_n{}^2\right)^{\frac{\gamma + 1}{2(1 - \gamma)}} \tag{6.27}$$

Multiplying both sides of Eq. (6.27) by $const = P_{ISA}/T_{ISA}$ we obtain

$$\frac{w_a\sqrt{\theta}}{A\delta} = const \sqrt{\frac{\gamma}{R}} M_n \left(1 + \frac{\gamma - 1}{2} M_n{}^2\right)^{\frac{\gamma + 1}{2(1 - \gamma)}} \tag{6.28}$$

Taking logs and differentiating

$$\begin{aligned}
\frac{dw_a}{w_a} + \frac{1}{2}\frac{d\theta}{\theta} - \frac{dA}{A} - \frac{d\delta}{\delta} &= \frac{1}{2}\left(\frac{d\gamma}{\gamma} - \frac{dR}{R}\right) + \frac{dM_n}{M_n} + \frac{\gamma + 1}{2(1 - \gamma)} \frac{d\left(1 + \frac{\gamma - 1}{2} M_n^2\right)}{1 + \frac{\gamma - 1}{2} M_n^2} \\
&= \frac{1}{2}\frac{d\gamma}{\gamma} + \frac{\gamma + 1}{2(1 - \gamma)} \left(\frac{\gamma M_n^2}{2\left(1 + \frac{\gamma - 1}{2} M_n^2\right)}\right) \frac{d\gamma}{\gamma} \\
&= \frac{1}{2} \underbrace{\left[1 + \frac{\gamma + 1}{2(1 - \gamma)} \frac{\gamma M_n^2}{\left(1 + \frac{\gamma - 1}{2} M_n^2\right)}\right]}_{x} \frac{d\gamma}{\gamma} \\
&= \frac{1}{2} x \frac{d\gamma}{\gamma}
\end{aligned} \tag{6.29}$$

In general, γ will depend on the local (station) temperature T. If (for approximation) we assume a form of

$$\gamma = \alpha T^{\beta} \Rightarrow \ln(\gamma) = \ln(\alpha) + \beta \ln(T) \tag{6.30}$$

From the plot in Fig. 6.1, we see that $\beta \approx -.041$. Therefore, we have

$$\frac{d\gamma}{\gamma} \approx -.041 \frac{dT}{T} = -.041 \alpha_T \frac{d\theta}{\theta} \tag{6.31}$$

which can be substituted in the relationship above to obtain

$$\frac{dw_a}{w_a}+\frac{1}{2}\frac{d\theta}{\theta}-\frac{dA}{A}-\frac{d\delta}{\delta}=\frac{1}{2}\underbrace{\left[1+\frac{\gamma+1}{2(1-\gamma)}\frac{\gamma M_n^2}{\left(1+\frac{\gamma-1}{2}M_n^2\right)}\right]}_{x}\left(-.041\ \alpha_T\frac{d\theta}{\theta}\right)$$
$$=\frac{1}{2}\left(-.041\ \ x\ \ \alpha_T\frac{d\theta}{\theta}\right) \tag{6.32}$$

Rearranging terms a little we have

$$\frac{dw_a}{w_a}+\frac{1}{2}(1+0.041x\alpha_T)\frac{d\theta}{\theta}-\frac{dA}{A}-\frac{d\delta}{\delta}=0 \tag{6.33}$$

Integrating this term provides

$$\frac{w_a}{A\delta\theta^{-1/2(1+0.041x\alpha_T)}}=0 \tag{6.34}$$

which indicates the appropriate correction

$$w_{ac}=\frac{w_a}{A\delta\theta^{-1/2(1+0.041x\alpha_T)}}=\frac{w_a}{A\delta\theta^{\alpha_{wa}}} \tag{6.35}$$

The value of $\boldsymbol{x}$ (in Eq. 6.35) will depend on M_n and T (since γ depends on T). Table 6.1 depicts the fractional change in airflow theta exponent values from classical (negative square root) for various temperatures and Mach number.

For core flow (typically at local Mach no of 0.3–0.4), an average α_{w_a} value would be in the range of approximately −0.505, which is close to the classical correction of −0.5.

6.4 Horsepower Correction

Horsepower required (or developed) by a compressor (or turbine) can be given by

$$\begin{aligned}&HP \propto (w_a\Delta h) \Rightarrow HP = const(w_a\Delta h)\\ \therefore\quad &HP = const(w_a cp\Delta T)\end{aligned} \tag{6.36}$$

where $\Delta T = T_{out} - T_{in}$ across the compressor (or turbine).

Employing our usual trick of taking logs and differentiating, we obtain the expression

Table 6.1 Fractional change in airflow theta exponent from classical

	Mach									
T (degK)	0.1	0.2	0.3	0.4	0.5	0.6	0.7	0.8	0.9	1
288.7	1.0196	1.0171	1.0129	1.0072	1.0000	0.9916	0.9821	0.9717	0.9606	0.9489
294.3	1.0196	1.0171	1.0129	1.0072	1.0000	0.9916	0.9821	0.9717	0.9606	0.9489
299.8	1.0196	1.0171	1.0129	1.0072	1.0000	0.9916	0.9821	0.9716	0.9605	0.9488
305.4	1.0196	1.0171	1.0129	1.0072	1.0000	0.9916	0.9821	0.9716	0.9605	0.9488
310.9	1.0196	1.0171	1.0129	1.0072	1.0000	0.9916	0.9821	0.9716	0.9605	0.9488
322.0	1.0196	1.0171	1.0129	1.0072	1.0000	0.9916	0.9821	0.9716	0.9605	0.9488
333.2	1.0196	1.0171	1.0129	1.0071	1.0000	0.9915	0.9820	0.9716	0.9604	0.9486
344.3	1.0196	1.0171	1.0129	1.0071	1.0000	0.9915	0.9820	0.9716	0.9604	0.9486
355.4	1.0196	1.0171	1.0129	1.0071	1.0000	0.9915	0.9820	0.9716	0.9604	0.9486
366.5	1.0196	1.0171	1.0129	1.0071	0.9999	0.9915	0.9820	0.9715	0.9603	0.9485
422.0	1.0196	1.0171	1.0128	1.0070	0.9998	0.9913	0.9817	0.9712	0.9599	0.9480
477.6	1.0196	1.0170	1.0128	1.0069	0.9997	0.9911	0.9814	0.9707	0.9593	0.9473
533.2	1.0196	1.0170	1.0127	1.0068	0.9995	0.9908	0.9810	0.9702	0.9587	0.9465
672.0	1.0196	1.0169	1.0125	1.0065	0.9989	0.9900	0.9799	0.9687	0.9568	0.9441
810.9	1.0196	1.0168	1.0123	1.0061	0.9983	0.9891	0.9787	0.9671	0.9547	0.9416
1088.7	1.0195	1.0167	1.0119	1.0055	0.9973	0.9877	0.9768	0.9647	0.9516	0.9377
1111.1	1.0195	1.0167	1.0119	1.0054	0.9973	0.9877	0.9767	0.9646	0.9515	0.9376
1388.9	1.0195	1.0166	1.0117	1.0050	0.9967	0.9867	0.9754	0.9629	0.9493	0.9349
1666.7	1.0195	1.0165	1.0115	1.0047	0.9962	0.9860	0.9745	0.9616	0.9477	0.9329
1944.4	1.0195	1.0164	1.0114	1.0045	0.9958	0.9855	0.9738	0.9607	0.9465	0.9314
2222.2	1.0195	1.0164	1.0113	1.0043	0.9956	0.9851	0.9732	0.9600	0.9456	0.9303
2500.0	1.0195	1.0164	1.0112	1.0042	0.9953	0.9848	0.9728	0.9594	0.9449	0.9294
2777.8	1.0195	1.0163	1.0112	1.0041	0.9951	0.9845	0.9724	0.9589	0.9443	0.9286

$$\frac{dHP}{HP} = \frac{dw_a}{w_a} + \frac{dcp}{cp} + \frac{d(\Delta T)}{\Delta T} = \frac{d\delta}{\delta} + \alpha_{w_a}\frac{d\theta}{\theta} + \frac{dcp}{cp} + \frac{d(\Delta T)}{\Delta T} \tag{6.37}$$

Expanding some terms

$$\frac{\Delta T}{2} = \frac{T_{out} - T_{in}}{2} = T_{average} \Rightarrow \frac{d(\Delta T)}{\Delta T} = \frac{d(\Delta T/2)}{\Delta T/2} = \frac{dT_{average}}{T_{average}} = \alpha_T \frac{d\theta}{\theta} \tag{6.38}$$

where α_T is the appropriate theta exponent for the $T_{average}$ under consideration. Combining expressions gives us

$$\begin{aligned}\frac{dHP}{HP} &= \frac{d\delta}{\delta} + \alpha_{w_a}\frac{d\theta}{\theta} + \frac{dcp}{cp} + \alpha_T\frac{d\theta}{\theta} \\ &= \frac{d\delta}{\delta} + \frac{dcp}{cp} + (\alpha_T + \alpha_{w_a})\frac{d\theta}{\theta}\end{aligned} \tag{6.39}$$

In general, the value of specific heat c_p will depend on local temperature as depicted in Fig. 6.4 below:

From the power function approximation, $cp \approx 0.1125\,T^{0.12}$, we have

$$\frac{dcp}{cp} \approx 0.12\frac{dT}{T} \tag{6.40}$$

Applying this to Eq. (6.39), it follows easily that

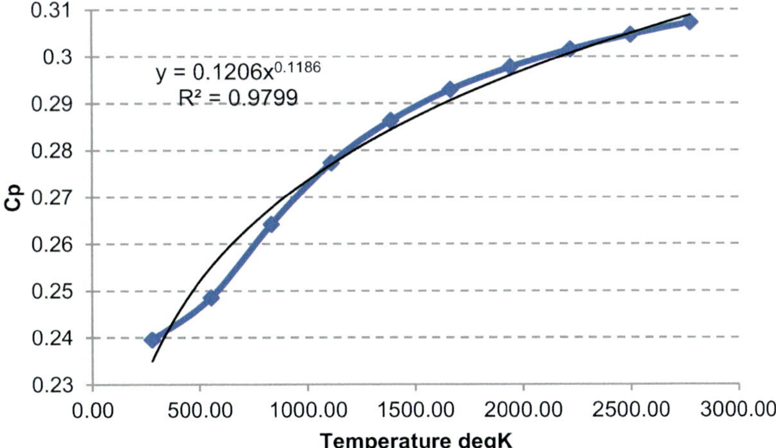

Fig. 6.4 Specific heat at constant pressure as a function of temperature

$$\begin{aligned}\frac{dHP}{HP} &= \frac{d\delta}{\delta} + \frac{dcp}{cp} + (\alpha_T + \alpha_{w_a})\frac{d\theta}{\theta} \\ &= \frac{d\delta}{\delta} + 0.12\frac{dT}{T} + (\alpha_T + \alpha_{w_a})\frac{d\theta}{\theta} = \frac{d\delta}{\delta} + 0.12\alpha_T\frac{d\theta}{\theta} + (\alpha_T + \alpha_{w_a})\frac{d\theta}{\theta} \\ &= \frac{d\delta}{\delta} + (1.12\alpha_T + \alpha_{w_a})\frac{d\theta}{\theta}\end{aligned} \tag{6.41}$$

Integrating yields the correction

$$HP_c = \frac{HP}{\delta\theta^{(1.12\alpha_T + \alpha_{wa})}} \tag{6.42}$$

6.5 Torque Correction

A new expression for corrected torque, Q, can be derived from the fundamental relationship between horsepower, torque and speed by leveraging the relationship in Eq. (6.41) as seen below.

$$\begin{aligned} HP &= QN \Rightarrow \\ \frac{dHP}{HP} &= \frac{dQ}{Q} + \frac{dN}{N} \Rightarrow \\ \frac{dQ}{Q} + \frac{dN}{N} - \frac{dHP}{HP} &= 0 \\ \frac{dQ}{Q} + 0.48\frac{d\theta}{\theta} - (1.12\alpha_T + \alpha_{w_a})\frac{d\theta}{\theta} - \frac{d\delta}{\delta} &= 0 \\ \frac{dQ}{Q} - (1.12\alpha_T + \alpha_{w_a} - 0.48)\frac{d\theta}{\theta} - \frac{d\delta}{\delta} &= 0 \end{aligned} \tag{6.43}$$

$$Q_c = \frac{Q}{\delta\theta^{1.12\alpha_T + \alpha_{wa} - 0.48}} \tag{6.44}$$

6.6 Acceleration Correction

Since $Q = I\dot{N}$, it follows immediately from Eq. (6.44) that

$$\dot{N}_c = \frac{\dot{N}}{\delta\theta^{1.12\alpha_T + \alpha_{wa} - 0.48}} \tag{6.45}$$

We end this chapter with a refinement to the classical correction for mass fuel flow which will follow from the previous results in this Chapter and formal manipulation.

6.7 Fuel Flow Correction

From basic energy balance we have

$$w_f \eta_b LHV = w_a \Delta h + w_f (h_4 - h_{wf}) \tag{6.46}$$

where $\Delta h = h_{out} - h_{in}$ is the enthalpy change across the combustor, i.e. $h_4 - h_3$. Combining and expanding, we have

$$\begin{aligned} w_f \eta_b LHV &= w_a \Delta h + w_f (h_4 - h_3 - h_{w_f} + h_3) \\ &= (w_a + w_f) \Delta h + w_f (h_4 - h_{w_f}) \\ &\approx (w_a + w_f) \Delta h \\ &\approx (w_a + w_f) cp \Delta T = (w_a + w_f) cp (T_3 - T_4) \end{aligned} \tag{6.47}$$

Taking logs of both sides and differentiating yields

$$\frac{dw_f}{w_f} = \frac{d(w_a cp \Delta T) + d(w_f cp \Delta T)}{w_a cp \Delta T + w_f cp \Delta T} = \frac{num}{den} \tag{6.48}$$

The two quantities in the numerator of Eq. (6.48) are found to be

$$\begin{aligned} d(w_a cp \Delta T) &= cp \Delta T dw_a + w_a (d(cp \Delta T)) = cp \Delta T dw_a + w_a (\Delta T\, dcp + cp\, d\Delta T) \\ &= cp \Delta T\, dw_a + w_a \Delta T\, dcp + w_a\, cp\, d\Delta T \end{aligned} \tag{6.49}$$

$$d(w_f cp \Delta T) = cp \Delta T\, dw_f + w_f \Delta T\, dcp + w_f cp\, d\Delta T \tag{6.50}$$

Inserting into Eqs. (6.49) and (6.50) into Eq. (6.48) and expanding

$$\begin{aligned} \frac{dw_f}{w_f} = \frac{num}{den} &= \frac{cp \Delta T (dw_a + dw_f) + (w_a + w_f) \Delta T\, dcp + (w_a + w_f) cp\, d\Delta T}{w_a cp\, \Delta T + w_f cp\, \Delta T} \\ &= \frac{(dw_a + dw_f) + (w_a + w_f) \Delta T \dfrac{dcp}{cp} + (w_a + w_f) \dfrac{d\Delta T}{\Delta T}}{w_a + w_f} \\ &= \frac{(dw_a + dw_f)}{w_a + w_f} + \frac{dcp}{cp} + \frac{d\Delta T}{\Delta T} \\ &= \left(\frac{w_a}{w_a + w_f} \right) \frac{dw_a}{w_a} + \left(\frac{w_f}{w_a + w_f} \right) \frac{dw_f}{w_f} + \frac{dcp}{cp} + \frac{d\Delta T}{\Delta T} \end{aligned} \tag{6.51}$$

where $\Delta T/2 = (T_4 - T_3)/2$ is the average temperature across the combustor. Rewriting in terms of the Fuel-Air-Ratio (FAR) we obtain

$$\begin{aligned} \frac{dw_f}{w_f} &= \left(\frac{1}{1+FAR}\right)\frac{dw_a}{w_a} + \left(\frac{FAR}{1+FAR}\right)\frac{dw_f}{w_f} + \frac{dcp}{cp} + \frac{d\Delta T}{\Delta T} \Rightarrow \\ \left(1 - \frac{FAR}{1+FAR}\right)\frac{dw_f}{w_f} &= \left(\frac{1}{1+FAR}\right)\frac{dw_a}{w_a} + \frac{dcp}{cp} + \frac{d\Delta T}{\Delta T} \Rightarrow \\ 0 &= \frac{dw_f}{w_f} - \frac{dw_a}{w_a} - (1+FAR)\frac{dcp}{cp} - (1+FAR)\frac{d\Delta T}{\Delta T} \end{aligned} \tag{6.52}$$

To simplify further, we apply Eq. (6.40) and other known quantities, namely

$$\begin{aligned} \frac{dcp}{cp} &\approx 0.12\frac{dT}{T} \\ \frac{dT}{T} &\approx \alpha_T\frac{d\theta}{\theta} \\ \frac{d\Delta T}{\Delta T} = \frac{d\Delta T/2}{\Delta T/2} = \frac{dT_{average}}{T_{average}} &\approx \alpha_T\frac{d\theta}{\theta} \end{aligned} \tag{6.53}$$

to obtain

$$\begin{aligned} \frac{dw_f}{w_f} - \frac{dw_a}{w_a} - (1+FAR)\frac{dcp}{cp} - (1+FAR)\frac{d\Delta T}{\Delta T} &= 0 \\ \frac{dw_f}{w_f} - \left(\frac{d\delta}{\delta} + \alpha_{w_a}\frac{d\theta}{\theta}\right) - (1+FAR)\left(1.12\alpha_T\frac{d\theta}{\theta}\right) &= 0 \\ \frac{dw_f}{w_f} - \frac{d\delta}{\delta} - \left[1.12\alpha_T(1+FAR) + \alpha_{w_a}\right]\frac{d\theta}{\theta} &= 0 \end{aligned} \tag{6.54}$$

The final expression in Eq. (6.54) provides the requisite correction:

$$w_{fc} = \frac{w_f}{\delta\theta^{1.12\alpha_T(1+FAR)+\alpha_{w_a}}} = \frac{w_f}{\delta\theta^{\alpha_{w_f}}} \tag{6.55}$$

For example, if FAR $= 0.0243$, and we assume classical correction for Temperature and Air Flow (i.e., $\alpha_T = 1$, $\alpha_{wa} = -0.5$) then

$$w_{fc} = \frac{w_f}{\delta\theta^{0.647}} \tag{6.56}$$

Chapter 7
Empirical Methods

Abstract This chapter considers empirical methodologies to determine (or refine) the θ and δ exponent values for use in Standard Day corrections. The section deviates from all previous chapters in that the corrections are not formally derived from axioms, assumptions and first principles but rather by statistical analysis of available engine data. The only consideration employed from our previous work is that of the *form* a corrected parameter takes, namely Eq. (2.3). Hypothetical data is employed to illustrate the techniques.

Keywords Empirical · Refinements · Corrections · Statistical analysis · Variation · Minimize · Reduce

In this Chapter, we will turn our attention to empirical methods that utilize actual engine data collected over an extended temperature (and pressure) range and apply statistical measures to determine the appropriate θ and δ exponents to normalize gas path parameters of interest. This data-centric consideration will be a departure from the previous Chapters in both philosophy as well as usage. At one extreme, we can think of this type of activity as a form of verification or fine tuning of the θ and δ exponents for various parameters whose (correction) forms were derived in the previous Chapters. Conversely, we could think of this exercise as a means to reduce variability in the data captured for analysis by allowing the θ and δ exponents to take on the form of *tuners* that are manipulated for the sole purpose of reducing variability and is independent of the interpretation of these as so called Standard Day Corrections. The empirical methods we will illustrate are the same in either event and are independent of interpretation. What is common is the notion of variance reduction. For example, it has become a generally accepted practice among engine performance diagnosticians to perform some sort of empirical analysis to verify the validity of the correction factors being used. Neglecting to do so exposes any subsequent diagnostic procedure to greater error which may render the diagnostic analysis suspect.

In terms of underlying assumptions we will borrow one result from Chap. 2, namely Eq. (2.3) as the form our normalization factor or correction will take, namely

A. J. Volponi, *Gas Turbine Parameter Corrections*,
https://doi.org/10.1007/978-3-030-41076-6_7

$$X_c \approx \frac{X}{\theta^a \delta^b} \quad where \quad \theta = \frac{T_2}{T_{ISA}}, \quad \delta = \frac{P_2}{P_{ISA}} \tag{7.1}$$

where X is the gas path parameter under consideration and the exponents a and b are the elements to be determined empirically. There are several possible techniques that can be put forward to analyze engine data for the purpose of normalization, however, we will restrict our present discussion to two popular (and simple) approaches. The first was derived by this author, circa 1980 and takes the form of a simple formula which we will re-derive here for completeness.

For simplicity, we will consider only temperature (θ) corrections; pressure (δ) corrections can be handled in an analogous manner. If we let X denote the gas path parameter under consideration (i.e., this could be a temperature, pressure, rotor speed, etc), we will assume we have available to us, a set of empirically observed data $\{X_i, \theta_i\}_{i=1}^n$ along with fundamental design information on what we would consider nominal levels for this parameter, $\left\{X_i^{(NOM)}\right\}_{i=1}^n$ to which we could compare the observational data in order to calculate a series of parameter residuals $\{\Delta X_i\}_{i=1}^n$. The nominal levels are typically calculated from a steady state engine model. Before proceeding, we need to clarify some implicit *assumptions* that we will impose on this process we are about to describe. The data $\{X_i, \theta_i\}_{i=1}^n$ can be from a single engine or a family of engines of the same model type. In the latter case, where more than one engine is represented, each engine must be at approximately the same performance deterioration level. Since we most likely will not know a given engine's health state, *a-priori*, we can qualify this assumption by insisting that the family consists of engines whose operating cycles (or hours) fall within a reasonably small band. The importance of this assumption will become clear in a moment.

We will assume that the parameter in question has a corrected form given by Eq. (7.1) and we let $X* = X/\theta^{\alpha_{True}}$, where α_{True} denotes the true (sought) exponent. We let α denote the theta exponent that will be used to normalize the observed data set $\{X_i, \theta_i\}_{i=1}^n$. We have that for each datum X_i (for all i), an inlet condition represented by θ_i as well as, the nominal value of this parameter, $X_i^{(NOM)}$, for the engine operating condition(s) at which the parameters X_i were obtained. With this information, we can calculate the % of point deviations as follows:

$$\Delta X_i = 100 \left(\frac{\frac{X_i}{\theta_i^\alpha} - \frac{X_i^{(NOM)}}{\theta_i^{\alpha_{True}}}}{\frac{X_i^{(NOM)}}{\theta_i^{\alpha_{True}}}} \right) \tag{7.2}$$

Rewriting this differently provides

$$\frac{X_i}{\theta_i^\alpha} = \frac{X_i^{(NOM)}}{\theta_i^{\alpha_{True}}} \left(1 + \frac{\Delta X_i}{100} \right) \tag{7.3}$$

Thus ΔX_i represents the % deviation (from nominal) and the *variation* we observe in these residuals are reflective of performance deterioration, sensor inaccuracies and inappropriate theta correction due to using an incorrect exponent α. Now if we plot these deltas against θ, then if α were the true correction exponent, we would expect to see a scatter plot that would have zero correlation, i.e. a linear regression line fit horizontal with the theta axis. An upward or downward trend, however, would indicate the extent to which the exponent α was deficient. If we let m and b, denote the regression slope and intercept respectively, then we have a defined linear relationship as follows:

$$\Delta X = m\theta + b \qquad \text{where } b = \Delta\overline{X} + m\overline{\theta} = \frac{1}{n}\sum_{i=1}^{n}\Delta X_i - m\left[\frac{1}{n}\sum_{i=1}^{n}\theta_i\right]$$

$$\text{and} \qquad m = r\frac{s_{\Delta X}}{s_\theta} \tag{7.4}$$

where r is the correlation coefficient and $s_{\Delta X}$ and s_θ are the sample standard deviations for ΔX and θ, respectively. Thus, for arbitrary θ, we have

$$\frac{X}{X^{(NOM)}}\theta^{\alpha_{True}-\alpha} \approx 1 + \frac{m\theta + b}{100} \Rightarrow \frac{X}{X^{(NOM)}} \approx 1 + \frac{m + b}{100} \quad (\theta = 1)$$

$$\text{and} \quad \frac{X}{X^{(NOM)}}\overline{\theta}^{\alpha_{True}-\alpha} \approx 1 + \frac{m\overline{\theta} + b}{100} \quad (\theta = \overline{\theta}) \tag{7.5}$$

therefore

$$\overline{\theta}^{\alpha_{True}-\alpha} \approx \left[1 + \frac{m\overline{\theta} + b}{100}\right] \Big/ \left[1 + \frac{m + b}{100}\right]$$

Taking natural logs of both sides and applying the relationship that $\ln(1 + x) \approx \boldsymbol{x}$ for small x, we obtain an expression for the "true" exponent in terms of the old exponent, namely,

$$\alpha_{True} \approx \alpha + \frac{m(\overline{\theta} - 1)}{100\ln(\overline{\theta})} \tag{7.6}$$

We see from this equation, that if the slope of the regression line is positive, then the exponent will be raised. Likewise, it will be lowered when the slope is negative. If the data sample was such that the mean theta was standard day (i.e., =1), then we cannot use Eq. (7.6) directly. However, since

$$\lim_{x \to 1} \left(\frac{x-1}{\ln x} \right) = 1 \tag{7.7}$$

Equation (7.6) reduces to a form we can utilize directly when $\overline{\theta} \approx 1$, i.e.

$$\alpha_{True} \approx \alpha + \frac{m}{100} \tag{7.8}$$

The form of Eqs. (7.6) and (7.8) are particularly simple and allows a rapid check for corrections when engine data is available. The reader should note that the above can be applied just as well to pressure corrections (i.e. the δ exponent) where the relationships would be identical except for replacing θ with δ).

At this juncture, it is important to note that there is a, perhaps not so obvious, nuance at play here. The only assumption we have made in this discussion is that of the form of the correction, namely Eq. (7.1). The reader may recall that this was derived through formal manipulation of our definition of a corrected quantity X_c as one that satisfies the following *implicit* relationship.

$$X = f(X_c, T_2, P_2) \tag{7.9}$$

In Chap. 2, we also observed that this relationship requires something being held constant and used local Mach number as the condition being held fixed. This assumption was used repeatedly in the Chaps. 2–6 in the formal derivations. Our usage in this empirical context, we have not made that formal assumption. In fact, it is plausible that what is actually at play here is the particular control paradigm in use and its power setting parameter (e.g., engine pressure ratio (EPR or corrected speed, $(N1/\sqrt{\theta})$, etc) being used to determine the reference condition $X^{(NOM)}$. The theta exponent derived, α_{True}, could be thought of as a tuning factor to force conformity of the observed data to the nominal model in terms of its relation to ambient condition (i.e., θ). As such, it becomes a *practical* correction factor as opposed to a *theoretical* correction factor. As such, this nuance is of little consequence to its usage in most real-world applications.

To illustrate Eqs. (7.6) and (7.8), we consider the following example. A hypothetical set of temperature data versus rotor speed is depicted below in Fig. 7.1. The assumed model nominal relationship is superimposed as the solid line appearing in the plot. Furthermore, each data point throughout was collected at a different ambient condition, i.e. varying levels of θ and a "*true*" theta exponent of 0.94 was assumed.

We follow the procedure described above wherein we first correct the data for the inlet condition and we do so in this example using an alpha = 0.8. We then calculate the % Δs and plot against θ as depicted in Fig. 7.2. A linear fit to the data has been superimposed in the plot.

The linear fit to this data has a slope $m = 13.8115$ and a mean theta of 0.9735. Using Eq. (7.6) we arrive at a new exponent of $\alpha = 0.936$. If we recalculate corrected

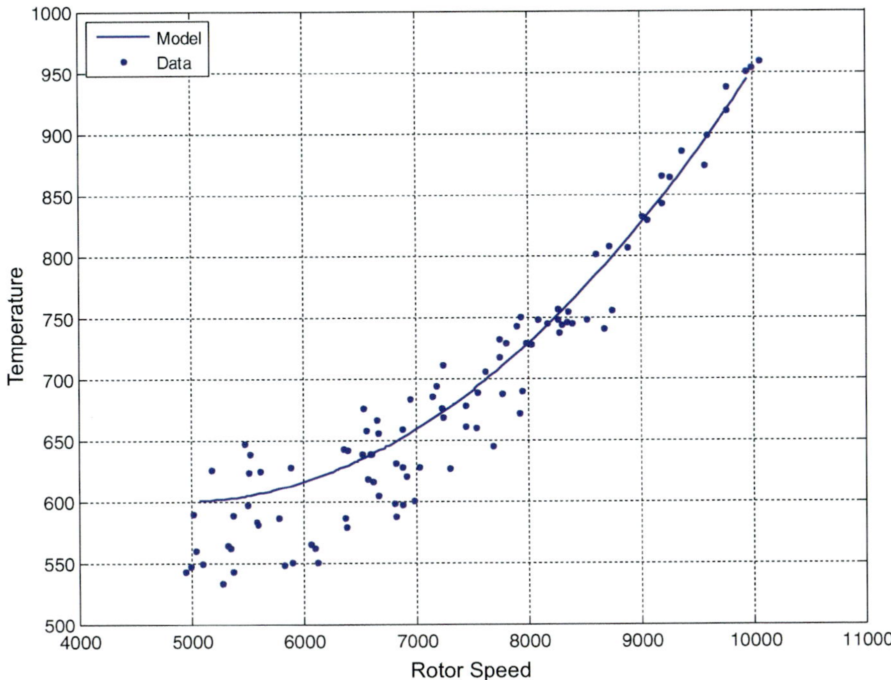

Fig. 7.1 Hypothetical temperature data vs. speed

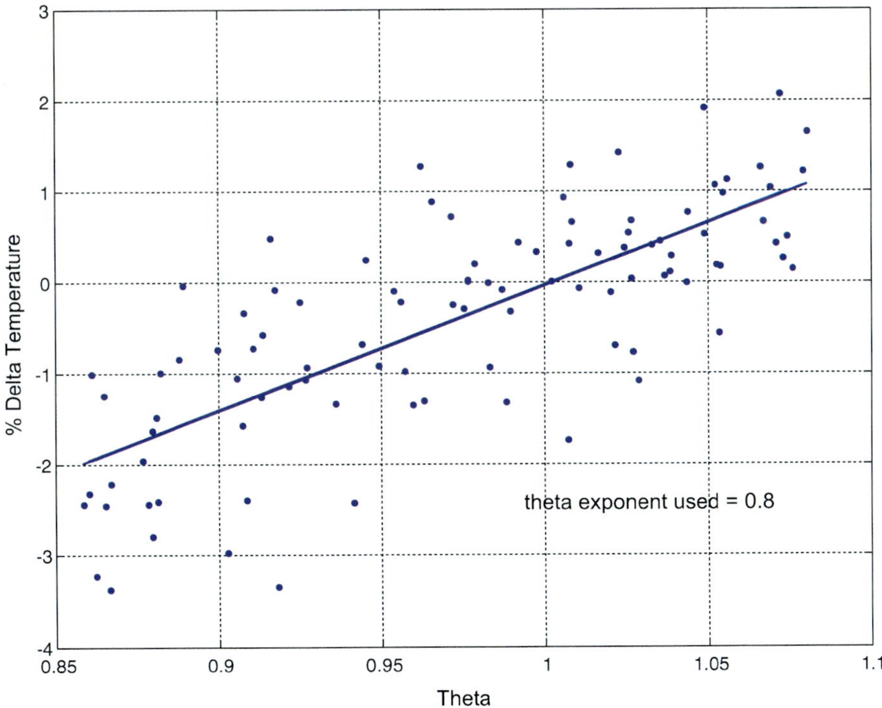

Fig. 7.2 % Temperature deltas vs. theta

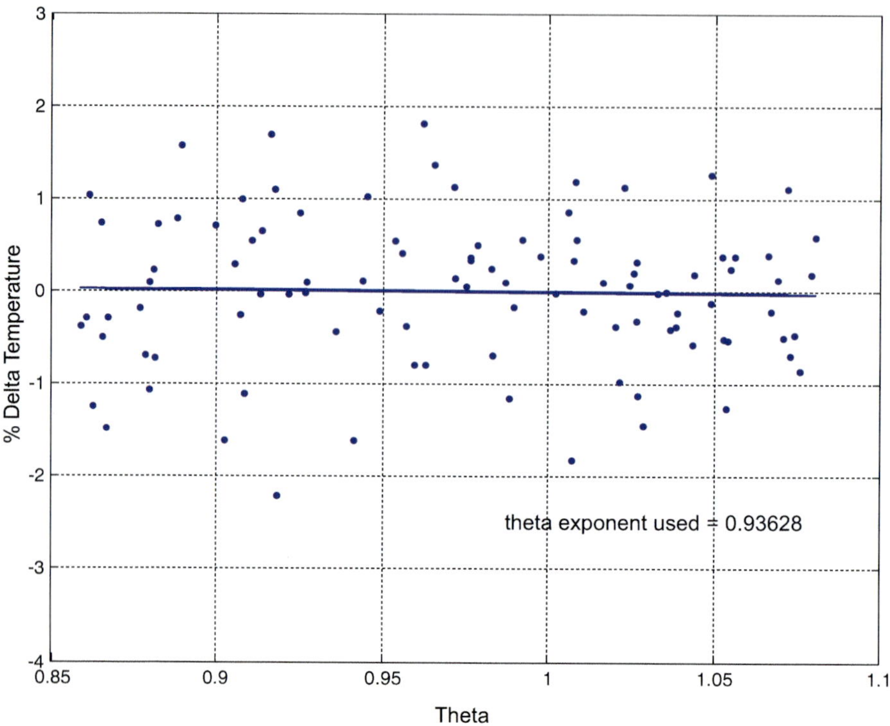

Fig. 7.3 % Temperature deltas vs. new theta

temperature using this new exponent and calculate the %Δs and plot them against theta (Fig. 7.3), we see that the slope is now essentially zero, indicating that there is now no correlation of the corrected parameter with ambient condition. From this we can conclude that the derived exponent $\alpha = 0.936$ fulfills the intent of a corrected parameter.

There are, of course, other empirical methods which one can employ in pursuit of the proper correction. For instance, one could determine the exponent which minimizes the standard deviation ($\sigma_{\Delta X}$) of the parameter deltas $\{\Delta X_i\}$, or to minimize the covariance ($\sigma_{\Delta X,\theta}$) between the deltas $\{\Delta X_i\}$ and $\{\theta_i\}$, or the correlation coefficient ($r_{\Delta X,\theta}$). Monte Carlo simulations indicate, however, that the results are approximately the same for all of these techniques. The differences obtained for theta exponents using these four methods are well within the limits of measurement noise and are not statistically significant. The three minimization methods which we have just alluded to, do carry a substantial computational overhead in that each of these methods are iterative in nature and require that a complete statistical analysis be performed at each newly chosen exponent, where the exponent is modulated in small increments over some range. After that is completed, the minimum statistic is isolated and the corresponding exponent is selected as the winner. To illustrate

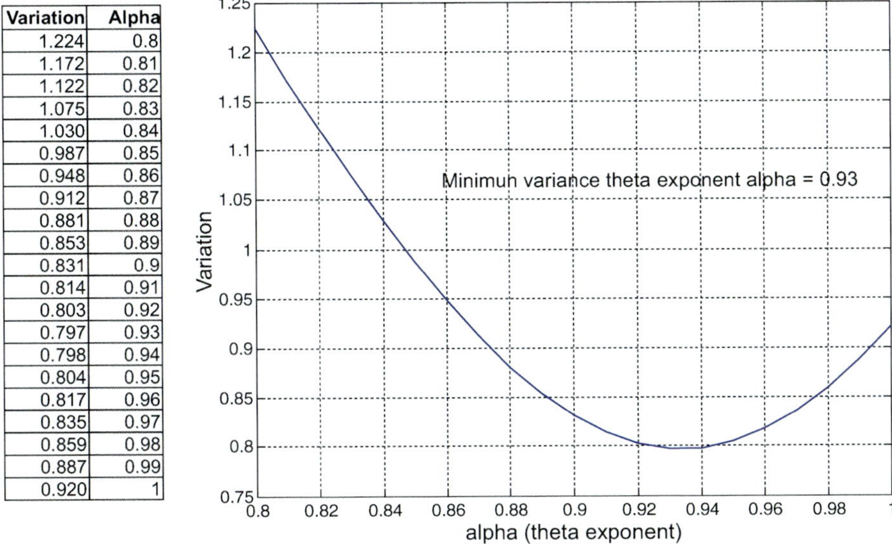

Variation	Alpha
1.224	0.8
1.172	0.81
1.122	0.82
1.075	0.83
1.030	0.84
0.987	0.85
0.948	0.86
0.912	0.87
0.881	0.88
0.853	0.89
0.831	0.9
0.814	0.91
0.803	0.92
0.797	0.93
0.798	0.94
0.804	0.95
0.817	0.96
0.835	0.97
0.859	0.98
0.887	0.99
0.920	1

Fig. 7.4 Effect of theta exponent α on parameter Δ variation

this method we will use the same data used above and repeat the analysis where we choose the exponent that will minimize the standard deviation ($\sigma_{\Delta X}$) of the parameter deltas.

To do this we pick a range of exponents $\alpha = [0.8, 0.81, 0.82, \ldots, 0.99, 1.0]$ from which to perform the analysis. For each exponent in this range, we re-compute the corrected Temperature, calculate its %Δ from nominal and calculate the variation (standard deviation, σ) present in the Δs. The result is presented in Fig. 7.4 above.

As can be seen from the graph, a value of 0.93 admits the minimum variance and is in complete alignment with the results of the previous methodology.

As with all empirically motivated methods, the nature of the data being utilized is of critical importance. The number of samples, the level of noise present, and the degree to which the data represents the effects under scrutiny are of critical importance to the outcome of the empirical method. In particular for this application we need to insure that our data represents a significant range of ambient conditions (i.e., θ). To do otherwise would severely compromise the analysis. To demonstrate this, we can repeat the previous analysis but reducing the ambient (temperature) range from [−15F, 100F] (or [−26C, 37.8C]) used earlier to a reduced range of [−15F, 0F] (or [−26C, −17.8C]. We will use the same starting point alpha = 0.8 to reduce the hypothetical data depicted in Figs. 7.5 and 7.6.

As can be seen from Fig. 7.6, the slope m is positive indicating that the α should be raised and using Eq. (7.2) yields a new estimate of $\alpha_{True} = 0.86$. This is in the right direction but falls short of the true value of 0.94. If we apply the second

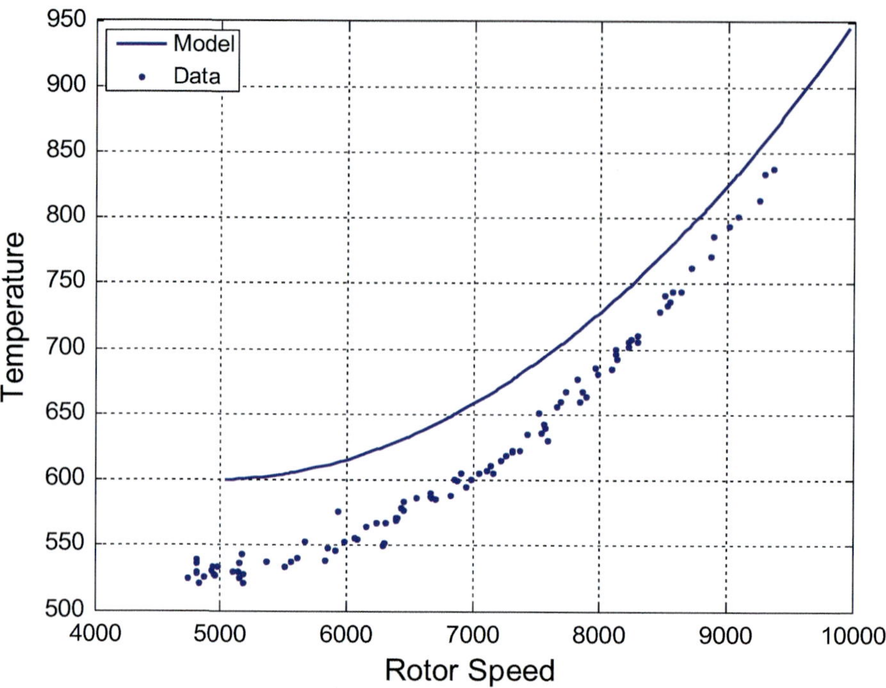

Fig. 7.5 Hypothetical temperature data vs. speed

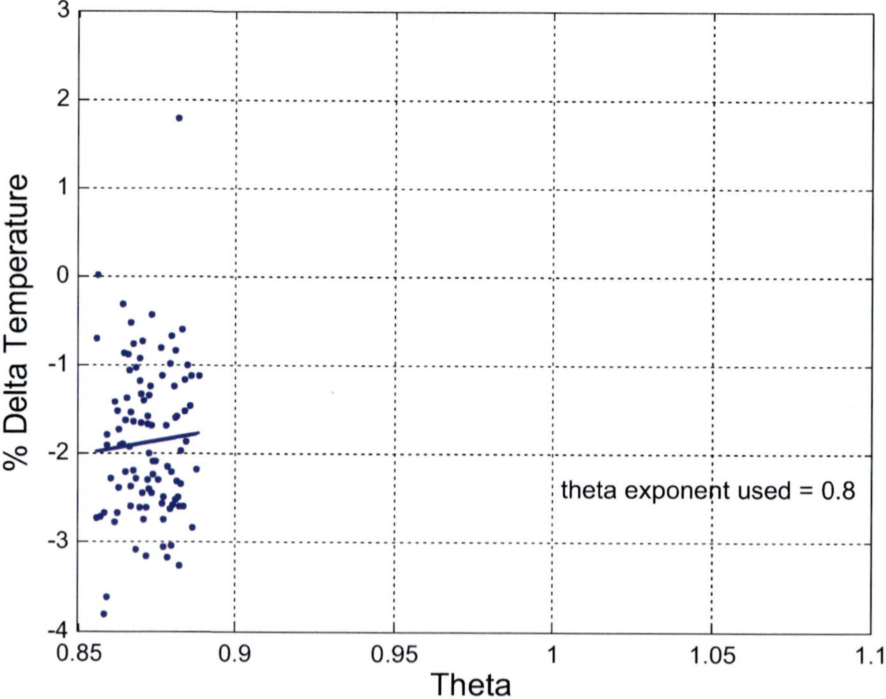

Fig. 7.6 % Temperature deltas vs. theta

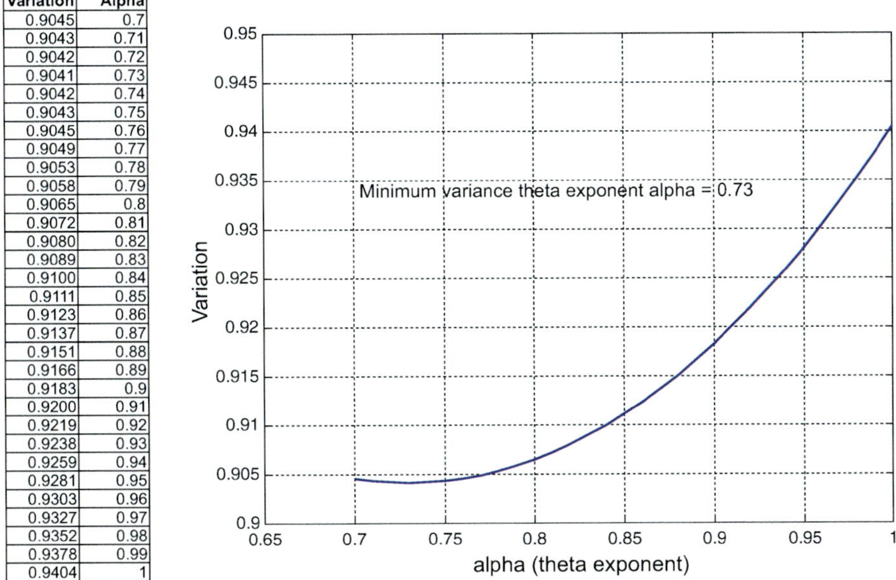

Variation	Alpha
0.9045	0.7
0.9043	0.71
0.9042	0.72
0.9041	0.73
0.9042	0.74
0.9043	0.75
0.9045	0.76
0.9049	0.77
0.9053	0.78
0.9058	0.79
0.9065	0.8
0.9072	0.81
0.9080	0.82
0.9089	0.83
0.9100	0.84
0.9111	0.85
0.9123	0.86
0.9137	0.87
0.9151	0.88
0.9166	0.89
0.9183	0.9
0.9200	0.91
0.9219	0.92
0.9238	0.93
0.9259	0.94
0.9281	0.95
0.9303	0.96
0.9327	0.97
0.9352	0.98
0.9378	0.99
0.9404	1

Fig. 7.7 Effect of theta exponent α on parameter Δ variation

methodology we obtain the information depicted in Fig. 7.7. This suggests a true alpha of 0.73! This not only misses the true value of 0.94 but is considerably different than the value predicted by the slope method. The caveat here is to make sure that data, *appropriate to task*, is collected before any analysis is performed and not to apply methodologies indiscriminately to an arbitrary collection of data.

Chapter 8
Humidity Corrections

Abstract It is well known that humidity levels in ambient air can affect engine performance. The intent of this chapter is to derive correction factors for gas path parameters of interest that will transform them to values that would have been observed had the ambient condition been *dry*. The strategy employed is to capture the effect of humidity on gas properties, namely, specific heats (cp and cv), their ratio (γ), and the gas constant (R), and extrapolate these levels to changes induced on gas path parameter values. Humidity corrections for Rotor Speed, Air Flow, Δ Temperature, Horsepower, Torque, Acceleration, Fuel Flow, Thrust, Pressure, and Temperature will be considered. Humidity corrections will be applied to Standard Day corrected parameters as an independent correction.

Keywords Specific humidity · Water-to-air ratio · Specific heats · Ratio of specific heats · Gas constant · Gas properties · Dry air · Wet air · Performance · Correction

8.1 Introduction

It has long been recognized that high humidity levels in the ambient air can affect engine performance, [8–12]. Since measured gas path quantities are used in evaluating engine performance deterioration, the omission of the effect of humidity essentially introduces increased uncertainty in these diagnostic calculations. Just as ambient (inlet) temperature and pressure are used to normalize gas path parameters before performance diagnostics are performed (standard day corrections), a similar normalization process is possible if the level of humidity is known. For the most part, humidity levels are not measured in aero-engines during flight, so much of this discourse has meaning and application to ground testing or for land-based gas turbines. The following analysis provides some insight to the magnitude of the effects encountered when humidity levels are present and measured. The analysis approach is to examine the effect humidity has on *gas properties*, in particular on four parameters, namely

A. J. Volponi, *Gas Turbine Parameter Corrections*,
https://doi.org/10.1007/978-3-030-41076-6_8

Specific heat at constant pressure	cp
Specific heat at constant volume	cv
Ratio of specific heats	γ
Gas constant	R

To set the stage for our derivations, we proceed with some preliminary definitions and relationships:

Define water to air ratio: $war = \frac{water\ \ mass}{dry\ \ air\ \ mass} = \frac{mass_w}{mass_{a,d}}$,

From this definition we can generate some additional relationships, i.e.

$$\begin{aligned} & total\ \ mass = (1 + war)mass_{a,d} \\ & SH = specific\ \ humidity = \frac{mass_w}{total\ \ mass} = \frac{mass_w}{(1 + war)mass_{a,d}} = \frac{war}{1 + war} \end{aligned} \tag{8.1}$$

Additionally, we define

$$\begin{aligned} cp_{a,d} &= \text{specific heat of dry air at constant pressure} \\ cp_w &= \text{specific heat of water (vapor) at constant pressure} \\ cp_{a,w} &= \text{specific heat of wet air at constant pressure} \end{aligned}$$

With a little manipulation we have that

$$\begin{aligned} cp_{a,w} &= \left(\frac{mass_{a,d}}{total\ \ mass}\right) cp_{a,d} + \left(\frac{mass_w}{total\ \ mass}\right) cp_w \\ &= \left(\frac{1}{1 + war}\right) cp_{a,d} + \left(\frac{war}{1 + war}\right) cp_w = (1 - SH)\, cp_{a,d} + SH\, cp_w \end{aligned} \tag{8.2}$$

Additionally

$$\begin{aligned} \frac{dcp}{cp} &= \frac{\Delta cp}{cp_{a,d}} \equiv \frac{cp_{a,w} - cp_{a,d}}{cp_{a,d}} = \frac{(1 - SH)cp_{a,d} + SH\, cp_w - cp_{a,d}}{cp_{a,d}} \\ &= \frac{SH\, cp_w - SH\, cp_{a,d}}{cp_{a,d}} \end{aligned} \tag{8.3}$$

Arriving at the expression

$$\frac{dcp}{cp} = SH\left(\frac{cp_w}{cp_{a,d}} - 1\right) \tag{8.4}$$

Equation (8.4) provides a fundamental relationship that we will leverage in developing the association between the specific heat of wet air ($cp_{a,\ w}$) and the specific heat of dry air ($cp_{a,\ d}$). Likewise, we develop analogous relationships for the gas constant R and the ratio of specific heats, γ. We define

$$R_{a,d} = \text{gas constant of dry air}$$
$$R_w = \text{gas constant of water (vapor)}$$
$$R_{a,w} = \text{gas constant of wet air}$$

Then

$$\begin{aligned} R_{a,w} &= \left(\frac{mass_{a,d}}{total\ mass}\right)R_{a,d} + \left(\frac{mass_w}{total\ mass}\right)R_w \\ &= (1 - SH)R_{a,d} + SH\,R_w \end{aligned} \tag{8.5}$$

Additionally

$$\begin{aligned} \frac{dR}{R} &= \frac{\Delta R}{R_{a,d}} \equiv \frac{R_{a,w} - R_{a,d}}{R_{a,d}} = \frac{(1 - SH)R_{a,d} + SH\,R_w - R_{a,d}}{R_{a,d}} \\ &= \frac{SH\,R_w - SH\,R_{a,d}}{R_{a,d}} \end{aligned} \tag{8.6}$$

Arriving at the expression

$$\frac{dR}{R} = SH\left(\frac{R_w}{R_{a,d}} - 1\right) \tag{8.7}$$

We need one more relationship involving the ratio of specific heats, $\gamma = cp/cv$. We start with a familiar relationship, i.e.

$$\begin{aligned} \frac{R}{cp} &= \frac{cp - cv}{cp} = 1 - \frac{cv}{cp} = 1 - \frac{1}{\gamma} \\ &= \frac{\gamma - 1}{k} \end{aligned} \tag{8.8}$$

Taking logs of both sides and differentiating we obtain

$$\begin{aligned} \frac{dR}{R} - \frac{dcp}{cp} &= \frac{d\gamma}{\gamma - 1} - \frac{d\gamma}{\gamma} = \left(\frac{\gamma}{\gamma - 1}\right)\frac{d\gamma}{\gamma} - \frac{d\gamma}{\gamma} = \left(\frac{\gamma}{\gamma - 1} - \frac{\gamma - 1}{\gamma - 1}\right)\frac{d\gamma}{\gamma} \\ &= \left(\frac{1}{\gamma - 1}\right)\frac{d\gamma}{\gamma} \end{aligned} \tag{8.9}$$

Yielding the final relationship

$$\frac{d\gamma}{\gamma} = (\gamma - 1)\left(\frac{dR}{R} - \frac{dcp}{cp}\right) \tag{8.10}$$

The following three expressions, Eqs. (8.4), (8.7), and (8.10) will be used throughout this Chapter.

$$\begin{aligned}\frac{dcp}{cp} &= SH\left(\frac{cp_w}{cp_{a,d}} - 1\right)\\ \frac{dR}{R} &= SH\left(\frac{R_w}{R_{a,d}} - 1\right)\\ \frac{d\gamma}{\gamma} &= (\gamma - 1)\left(\frac{dR}{R} - \frac{dcp}{cp}\right)\end{aligned} \tag{8.11}$$

Furthermore, if we evaluate the expressions in Eq. (8.11) using the following standard values for these gas parameters, i.e.

$$\begin{aligned}cp_{a,d} &= 1004 \;\; J/kgDegK\\ cp_w &= 1859 \;\; J/kgDegK\\ R_{a,d} &= 287 \;\; J/kgDegK\\ R_w &= 461 \;\; J/kgDegK\end{aligned}$$

we have the following (approximate) relationships:

$$\begin{aligned}\frac{dcp}{cp} &\approx 0.852\left(\frac{war}{1+war}\right) = 0.852\,SH\\ \frac{dR}{R} &\approx 0.61\left(\frac{war}{1+war}\right) = 0.61\,SH\\ \frac{d\gamma}{\gamma} &\approx -.0968\left(\frac{war}{1+war}\right) = -0.0968\,SH\end{aligned} \tag{8.12}$$

Now, in general, for an arbitrary gas path parameter *X*, we can write the following:

$$\frac{X_{a,w}}{X_{a,d}} = 1 + \frac{X_{a,w}}{X_{a,d}} - \frac{X_{a,d}}{X_{a,d}} = 1 + \frac{dX}{X} \Rightarrow \frac{X_{a,d}}{X_{a,w}} = \frac{1}{1 + \frac{dX}{X}} \tag{8.13}$$

Thus, letting $X = cp$, R, and γ, we have the general relationships, which we will apply throughout our derivations to follow.

$$\begin{aligned}\frac{\gamma_{a,d}}{\gamma_{a,w}} &\approx \frac{1}{[1 - 0.0968 \;\; SH]}\\ \frac{R_{a,d}}{R_{a,w}} &\approx \frac{1}{[1 + 0.61 \;\; SH]}\\ \frac{cp_{a,d}}{cp_{a,w}} &\approx \frac{1}{[1 + 0.852 \;\; SH]}\end{aligned} \tag{8.14}$$

The relationships given in Eq. (8.14) will be used to derive a humidity correction for various gas path parameters of interest. This correction factor, which we will denote by μ_X, can be applied to a parameter X_{wet} under "*wet*" conditions to infer what its *equivalent value* would be under "*dry*" conditions as follows:

$$X_{dry} = X_{wet}\mu_X \tag{8.15}$$

The form and values for μ_X will be combinations of the terms appearing in Eq. (8.14). We begin by considering spool speed as way of example.

8.2 Spool Speed Humidity Correction

We know, in general, that

$$N = \frac{M_n}{r}\sqrt{g\gamma RT} \tag{8.16}$$

With a little manipulation (and assuming M_n constant) we obtain

$$\left(\frac{N}{\sqrt{T}}\right)_{a,w} = \left(\frac{N}{\sqrt{T}}\right)_{a,d}\sqrt{\frac{\gamma_{a,w}}{\gamma_{a,d}}\frac{R_{a,w}}{R_{a,d}}} \tag{8.17}$$

Manipulating this last expression and leveraging Eq. (8.14), we obtain the correction

$$\left(\frac{N}{\sqrt{T}}\right)_{a,d} = \left(\frac{N}{\sqrt{T}}\right)_{a,w}\frac{1}{\sqrt{[1-0.0968\ SH][1+0.61\ SH]}} = \left(\frac{N}{\sqrt{T}}\right)_{a,w}\mu_N \tag{8.18}$$

which is equivalent to

$$\left(\frac{N}{\sqrt{\theta}}\right)_{a,d} = \left(\frac{N}{\sqrt{\theta}}\right)_{a,w}\frac{1}{\sqrt{[1-0.0968\ SH][1+0.61\ SH]}} = \left(\frac{N}{\sqrt{\theta}}\right)_{a,w}\mu_N \tag{8.19}$$

8.3 Air Flow Humidity Correction

We start with a familiar relationship between corrected airflow and Mach number.

$$\frac{w_a\sqrt{T}}{AP} = M_n\sqrt{\frac{g\gamma}{R}}\left(1+\frac{\gamma-1}{2}M_n{}^2\right)^{\frac{\gamma+1}{2(1-\gamma)}} \tag{8.20}$$

Thus we can form the ratio

$$\frac{\left(\frac{w_a\sqrt{T}}{AP}\right)_{a,w}}{\left(\frac{w_a\sqrt{T}}{AP}\right)_{a,d}} = \frac{\sqrt{\frac{\gamma_{a,w}}{R_{a,w}}}}{\sqrt{\frac{\gamma_{a,d}}{R_{a,d}}}}\left[\frac{\left(1+\frac{\gamma_{a,w}-1}{2}M_n{}^2\right)^{\frac{\gamma_{a,w}+1}{2(1-\gamma_{a,w})}}}{\left(1+\frac{\gamma_{a,d}-1}{2}M_n{}^2\right)^{\frac{\gamma_{a,d}+1}{2(1-\gamma_{a,d})}}}\right]$$

$$= \sqrt{\frac{\frac{\gamma_{a,w}}{\gamma_{a,d}}}{\frac{R_{a,w}}{R_{a,d}}}}\left[\frac{\left(1+\frac{\gamma_{a,w}-1}{2}M_n{}^2\right)^{\frac{\gamma_{a,w}+1}{2(1-\gamma_{a,w})}}}{\left(1+\frac{\gamma_{a,d}-1}{2}M_n{}^2\right)^{\frac{\gamma_{a,d}+1}{2(1-\gamma_{a,d})}}}\right] \quad (8.21)$$

Applying the values from Eq. (8.14) we obtain

$$\frac{\left(\frac{w_a\sqrt{T}}{AP}\right)_{a,w}}{\left(\frac{w_a\sqrt{T}}{AP}\right)_{a,d}} = \sqrt{\frac{1-0.096SH}{1+0.61SH}}\left[\frac{\left(1+\frac{\gamma_{a,w}-1}{2}M_n{}^2\right)^{\frac{\gamma_{a,w}+1}{2(1-\gamma_{a,w})}}}{\left(1+\frac{\gamma_{a,d}-1}{2}M_n{}^2\right)^{\frac{\gamma_{a,d}+1}{2(1-\gamma_{a,d})}}}\right] \quad (8.22)$$

Looking at the last term in brackets in Eq. (8.22), and rewriting the *numerator* as

$$\left(1+\frac{\gamma_{a,w}-1}{2}M_n{}^2\right)^{\frac{\gamma_{a,w}+1}{2(1-\gamma_{a,w})}} = \left(1+\frac{\frac{\gamma_{a,w}}{\gamma_{a,d}}-\frac{1}{\gamma_{a,d}}}{\frac{2}{\gamma_{a,d}}}M_n{}^2\right)^{\frac{\frac{\gamma_{a,w}}{\gamma_{a,d}}+\frac{1}{\gamma_{a,d}}}{2\left(\frac{1}{\gamma_{a,d}}-\frac{\gamma_{a,w}}{\gamma_{a,d}}\right)}}$$

$$= \left(1+\frac{[1-0.0968SH]-\frac{1}{\gamma_{a,d}}}{\frac{2}{\gamma_{a,d}}}M_n{}^2\right)^{\frac{[1-0.0968SH]+\frac{1}{\gamma_{a,d}}}{2\left(\frac{1}{\gamma_{a,d}}-[1-0.0968SH]\right)}} \quad (8.23)$$

Substituting this expression into Eq. (8.22), we see that the *bracketed* expression becomes

$$\left[\frac{\left(1+\frac{\gamma_{a,w}-1}{2}M_n{}^2\right)^{\frac{\gamma_{a,w}+1}{2(1-\gamma_{a,w})}}}{\left(1+\frac{\gamma_{a,d}-1}{2}M_n{}^2\right)^{\frac{\gamma_{a,d}+1}{2(1-\gamma_{a,d})}}}\right] = \frac{\left(1+\frac{[1-0.0968SH]-\frac{1}{\gamma_{a,d}}}{\frac{2}{\gamma_{a,d}}}M_n{}^2\right)^{\frac{[1-0.0968SH]+\frac{1}{\gamma_{a,d}}}{2\left(\frac{1}{\gamma_{a,d}}-[1-0.0968SH]\right)}}}{\left(1+\frac{\gamma_{a,d}-1}{2}M_n{}^2\right)^{\frac{\gamma_{a,d}+1}{2(1-\gamma_{a,d})}}} \quad (8.24)$$

Evaluating the Mach no expression on the right hand side of the foregoing expression for various values of flow Mach No and specific humidity (SH), we see that from the Table 8.1 below that *error* incurred by replacing this Mach No factor with unity is acceptably small.

Table 8.1 Mach No. error

	Flow Mach No (%)									
SH	0.1	0.2	0.3	0.4	0.5	0.6	0.7	0.8	0.9	1
0.00000	0.00	0.00	0.00	0.00	0.00	0.00	0.00	0.00	0.00	0.00
0.01000	0.00	−0.01	−0.02	−0.03	−0.05	−0.07	−0.09	−0.12	−0.14	−0.17
0.02000	0.00	−0.02	−0.04	−0.06	−0.10	−0.14	−0.18	−0.23	−0.29	−0.34
0.03000	−0.01	−0.02	−0.05	−0.10	−0.15	−0.21	−0.27	−0.35	−0.43	−0.51
0.04000	−0.01	−0.03	−0.07	−0.13	−0.20	−0.28	−0.37	−0.46	−0.57	−0.68

Thus, the correction for air flow becomes

$$\left(\frac{w_a\sqrt{\theta}}{A\delta}\right)_{a,d} = \left(\frac{w_a\sqrt{\theta}}{A\delta}\right)_{a,w}\left(\sqrt{\frac{1+0.61SH}{1-.0968SH}}\right) = \left(\frac{w_a\sqrt{\theta}}{A\delta}\right)_{a,w}\mu_{w_a} \tag{8.25}$$

8.4 ΔTemperature Humidity Correction

8.4.1 *ΔTotal − Static Temperature Correction*

Given that

$$\frac{T}{T_s} = \left(1+\frac{\gamma-1}{2}M_n^2\right) \quad and \quad \gamma = cp/cv, R = cp - cv \Rightarrow \frac{\gamma-1}{2} = \frac{R\gamma}{2cp} \tag{8.26}$$

Rewriting the expression yields

$$\left\{\left(\frac{T}{T_s}\right) - 1\right\} = \frac{R\gamma}{cp}\frac{1}{2}M_n^2 \tag{8.27}$$

Forming the dry to wet ratios and letting $\Delta T = T - T_s$ provides the following expression.

$$\frac{\left(\frac{T}{T_s}\right)_{a,d} - 1}{\left(\frac{T}{T_s}\right)_{a,w} - 1} = \frac{\left(\frac{\Delta T}{T_s}\right)_{a,d}}{\left(\frac{\Delta T}{T_s}\right)_{a,w}} = \frac{R_{a,d}\gamma_{a,d}}{R_{a,w}\gamma_{a,w}}\frac{cp_{a,w}}{cp_{a,d}} = \frac{(1+0.852SH)}{(1+0.61SH)(1-0.0968SH)}$$
$$\equiv \mu_{\Delta T/T_s} \tag{8.28}$$

Rewriting, we have the correction for $\Delta T/T_s$

$$\left(\frac{\Delta T}{T_s}\right)_{a,d} = \frac{(1+0.852SH)}{(1+0.61SH)(1-0.0968SH)}\left(\frac{\Delta T}{T_s}\right)_{a,w} \tag{8.29}$$

8.4.2 Δ Enthalpy Correction

We start with a fundamental relationship between enthalpy (h) and internal energy (U) as follows

$$\Delta h = \Delta U + P\Delta V \tag{8.30}$$

If we assume that the volume is fixed ($\Delta V = 0$) then the change in enthalpy is equivalent to the change in internal energy. If we represent this change in kinetic terms, we have

$$\Delta h = \frac{1}{2}mv^2 = \frac{1}{2}mM_n{}^2 g\gamma RT \tag{8.31}$$

from which it follows that

$$\frac{\Delta h}{T} = \frac{1}{2}mM_n{}^2 g\gamma R \tag{8.32}$$

Forming the wet versus dry ratio from this relationship and applying Eq. (8.14) yields

$$\frac{\left(\frac{\Delta h}{T}\right)_{a,d}}{\left(\frac{\Delta h}{T}\right)_{a,w}} = \frac{\gamma_{a,d}R_{a,d}}{\gamma_{a,w}R_{a,w}} = \frac{1}{(1-0.0968SH)(1+0.61SH)} \equiv \mu_{\Delta h} \tag{8.33}$$

Therefore

$$\left(\frac{\Delta h}{T}\right)_{a,d} = \frac{1}{(1-0.0968SH)(1+0.61SH)}\left(\frac{\Delta h}{T}\right)_{a,w} \tag{8.34}$$

8.4.3 Δ Temperature Correction

We can approximate a change in enthalpy by $\Delta h = cp\,\Delta T$, therefore

$$\frac{\Delta T}{T} = \frac{\Delta h}{cpT} \Rightarrow$$

$$\frac{\left(\frac{\Delta T}{T}\right)_{a,d}}{\left(\frac{\Delta T}{T}\right)_{a,w}} = \frac{\left(\frac{\Delta h}{T}\right)_{a,d}\frac{1}{cp_{a,d}}}{\left(\frac{\Delta h}{T}\right)_{a,w}\frac{1}{cp_{a,w}}} = \frac{cp_{a,w}}{cp_{a,d}}\mu_{\Delta h} = \frac{(1+0.852SH)}{(1-0.0968SH)(1+0.61SH)} \equiv \mu_{\Delta T/T} \tag{8.35}$$

This provides the correction

$$\left(\frac{\Delta T}{T}\right)_{a,d} = \frac{(1+0.852SH)}{(1-0.0968SH)(1+0.61SH)}\left(\frac{\Delta T}{T}\right)_{a,w} \tag{8.36}$$

8.5 Horsepower Humidity Correction

Starting with definition

$$HP = \frac{J}{550} wcp\Delta T \tag{8.37}$$

Dividing both sides of Eq. (8.30) by $\sqrt{T}$and P, we obtain

$$\frac{HP}{\sqrt{T}P} = \frac{J}{550}\frac{w\sqrt{T}}{P}cp\frac{\Delta T}{T} \tag{8.38}$$

Applying Eqs. (8.14), (8.25) and (8.36)

$$\begin{aligned}\frac{\left(\frac{HP}{\sqrt{T}P}\right)_{a,d}}{\left(\frac{HP}{\sqrt{T}P}\right)_{a,w}} &= \frac{\frac{J}{550}\left(\frac{w_a\sqrt{T}}{AP}\right)_{a,d} cp_{a,d}\left(\frac{\Delta T}{T}\right)_{a,d}}{\frac{J}{550}\left(\frac{w_a\sqrt{T}}{AP}\right)_{a,w} cp_{a,w}\left(\frac{\Delta T}{T}\right)_{a,w}} = \frac{cp_{a,d}}{cp_{a,w}}\mu_{w_a}\mu_{\Delta T/T} \\ &= \frac{1}{(1+0.852SH)}\left(\sqrt{\frac{1+0.61SH}{1-.0968SH}}\right)\left(\frac{(1+0.852SH)}{(1-0.0968SH)(1+0.61SH)}\right) \\ &= \frac{1}{\sqrt{(1+0.61SH)}(1-0.0968SH)^{1.5}}\end{aligned} \tag{8.39}$$

This yields the correction

$$\begin{aligned}\left(\frac{HP}{\sqrt{\theta}\delta}\right)_{a,d} &= \left(\frac{HP}{\sqrt{\theta}\delta}\right)_{a,w}\frac{1}{\sqrt{(1+0.61SH)}(1-0.0968SH)^{1.5}} \\ &= \left(\frac{HP}{\sqrt{\theta}\delta}\right)_{a,w}\mu_{HP}\end{aligned} \tag{8.40}$$

8.6 Torque Humidity Correction

Starting with the relationship between horsepower, torque, and rotational speed, we have

$$HP = QN \Rightarrow \frac{HP}{\sqrt{T}P} = \frac{Q}{P}\frac{N}{\sqrt{T}} \Rightarrow$$

$$\frac{\left(\frac{HP}{\sqrt{T}P}\right)_{a,d}}{\left(\frac{HP}{\sqrt{T}P}\right)_{a,w}} = \frac{\left(\frac{Q}{P}\right)_{a,d}\left(\frac{N}{\sqrt{T}}\right)_{a,d}}{\left(\frac{Q}{P}\right)_{a,w}\left(\frac{N}{\sqrt{T}}\right)_{a,w}} \Rightarrow \frac{\left(\frac{Q}{P}\right)_{a,d}}{\left(\frac{Q}{P}\right)_{a,w}} = \frac{\left(\frac{HP}{\sqrt{T}P}\right)_{a,d}}{\left(\frac{HP}{\sqrt{T}P}\right)_{a,w}}\frac{\left(\frac{N}{\sqrt{T}}\right)_{a,w}}{\left(\frac{N}{\sqrt{T}}\right)_{a,d}} = \frac{\mu_{HP}}{\mu_N} \tag{8.41}$$

Applying Eqs. (8.14), (8.19), and (8.40), we obtain

$$\left(\frac{Q}{\delta}\right)_{a,d} = \left(\frac{Q}{\delta}\right)_{a,w}\left[\frac{1}{\sqrt{(1+0.61SH)(1-0.0968SH)^{1.5}}}\right] \times \sqrt{[1-0.0968SH][1+0.61SH]} \tag{8.42}$$

After some consolidation we obtain the correction

$$\left(\frac{Q}{\delta}\right)_{a,d} = \left(\frac{Q}{\delta}\right)_{a,w}\left[\frac{1}{(1-0.0968SH)}\right] = \left(\frac{Q}{\delta}\right)_{a,w}\mu_Q \tag{8.43}$$

8.7 Acceleration Humidity Correction

Since

$$Q = I\dot{N} \tag{8.44}$$

we have

$$\frac{\left(\frac{\dot{N}}{\delta}\right)_{a,d}}{\left(\frac{\dot{N}}{\delta}\right)_{a,w}} = \frac{\left(\frac{Q}{\delta}\right)_{a,d}}{\left(\frac{Q}{\delta}\right)_{a,w}} = \mu_Q \tag{8.45}$$

yielding

$$\left(\frac{\dot{N}}{\delta}\right)_{a,d} = \left(\frac{\dot{N}}{\delta}\right)_{a,w}\left[\frac{1}{(1-0.0968SH)}\right] = \left(\frac{\dot{N}}{\delta}\right)_{a,w}\mu_{\dot{N}} \tag{8.46}$$

8.8 Fuel Flow Humidity Correction

Starting with the familiar energy equation bounding the combustor

$$w_f \eta_b LHV = w_g \Delta h = \left(w_a + w_f\right) \Delta h \Rightarrow$$
$$\frac{w_f}{\sqrt{\theta\delta}} \eta_b LHV = \frac{w_g \sqrt{\theta}}{\delta} \frac{\Delta h}{\theta} = \frac{\left(w_a + w_f\right)\sqrt{\theta}}{\delta} \frac{\Delta h}{\theta} \approx \left(\frac{w_a \sqrt{\theta}}{\delta} \frac{\Delta h}{\theta}\right) \Rightarrow \tag{8.47}$$

Forming the usual dry to wet ratio, we obtain

$$\frac{\left(\frac{w_f}{\sqrt{\theta\delta}}\right)_{a,d}}{\left(\frac{w_f}{\sqrt{\theta\delta}}\right)_{a,w}} \frac{(\eta_b LHV)}{(\eta_b LHV)} = \frac{\left(\frac{w_a \sqrt{\theta}}{\delta}\right)_{a,d}}{\left(\frac{w_a \sqrt{\theta}}{\delta}\right)_{a,w}} \frac{\left(\frac{\Delta h}{\theta}\right)_{a,d}}{\left(\frac{\Delta h}{\theta}\right)_{a,w}} \tag{8.48}$$

After some cancellation of terms and applying preceding results, we see that

$$\frac{\left(\frac{w_f}{\sqrt{\theta\delta}}\right)_{a,d}}{\left(\frac{w_f}{\sqrt{\theta\delta}}\right)_{a,w}} = \mu_w \mu_{\Delta h} = \frac{1}{(1 - 0.0968SH)(1 + 0.61SH)} \left(\sqrt{\frac{1 + 0.61SH}{1 - .0968SH}}\right) \tag{8.49}$$

Providing the correction

$$\begin{aligned} \left(\frac{w_f}{\sqrt{\theta\delta}}\right)_{a,d} &= \left(\frac{w_f}{\sqrt{\theta\delta}}\right)_{a,w} \left\{\frac{1}{(1 - 0.0968SH)^{1.5}\sqrt{(1 + 0.61SH)}}\right\} \\ &= \left(\frac{w_f}{\sqrt{\theta\delta}}\right)_{a,w} \mu_{w_f} \end{aligned} \tag{8.50}$$

8.9 Thrust Humidity Correction (Turbojet)

We start with the basic definition of net thrust, F_n, used in Chap. 3

$$F_n = \frac{w_9}{g} v_9 - \frac{w_2}{g} v_2 + A_9(P_9 - P_{s9}) \tag{8.51}$$

where station 9 is assumed to be at the engine exhaust nozzle exit. We can simplify this expression by considering an equivalent over-expansion by determining a $v_{9expanded}$ such that

$$\frac{w_9}{g} v_{9expanded} = A_9(P_9 - P_{s9}) \Rightarrow F_n = \frac{w_9}{g} v_9 - \frac{w_2}{g} v_2 + \frac{w_9}{g} v_{9expanded}$$
$$= \frac{w_9}{g} v_9^* - \frac{w_2}{g} v_2 \tag{8.52}$$

where $v_9^* = v_9 + v_{9expanded}$. Therefore

$$\begin{aligned} F_n &= \frac{w_9}{g} v_9^* - \frac{w_2}{g} v_2 \\ &= \frac{w_9}{g} \sqrt{2gJ\Delta h_9^*} - \frac{w_2}{g} \sqrt{2gJ\Delta h_2} \quad where \quad \Delta h = h_T - h_S \end{aligned} \tag{8.53}$$

Since $w_9 = w_2 + w_f = w_2(1 + FAR) \approx w_2$, we have

$$F_n \approx w_9 \sqrt{\frac{2J}{g}} \left(\sqrt{\Delta h_9^*} - \sqrt{\Delta h_2} \right) \tag{8.54}$$

Thus

$$\frac{F_n}{\delta} \approx \sqrt{\frac{2J}{g}} \frac{w_9 \sqrt{\theta}}{\delta} \left(\sqrt{\frac{\Delta h_9^*}{\theta}} - \sqrt{\frac{\Delta h_2}{\theta}} \right) \tag{8.55}$$

Thus, the ratio of wet and dry corrected quantities yield

$$\begin{aligned} \frac{\left(\frac{F_n}{\delta}\right)_{a,d}}{\left(\frac{F_n}{\delta}\right)_{a,w}} &= \frac{\left(\frac{w_9\sqrt{\theta}}{\delta}\right)_{a,d}}{\left(\frac{w_9\sqrt{\theta}}{\delta}\right)_{a,w}} \frac{\sqrt{\left(\frac{\Delta h_9^*}{\theta}\right)_{a,d}} - \sqrt{\left(\frac{\Delta h_2}{\theta}\right)_{a,d}}}{\sqrt{\left(\frac{\Delta h_9^*}{\theta}\right)_{a,w}} - \sqrt{\left(\frac{\Delta h_2}{\theta}\right)_{a,w}}} \\ &= \mu_w \frac{\sqrt{\left(\frac{\Delta h_9^*}{\theta}\right)_{a,d}} - \sqrt{\left(\frac{\Delta h_2}{\theta}\right)_{a,d}}}{\sqrt{\left(\frac{\Delta h_9^*}{\theta}\right)_{a,w}} - \sqrt{\left(\frac{\Delta h_2}{\theta}\right)_{a,w}}} \end{aligned} \tag{8.56}$$

In general, we know from previous calculations that

$$\frac{\left(\frac{\Delta h}{T}\right)_{a,d}}{\left(\frac{\Delta h}{T}\right)_{a,w}} = \frac{k_{a,d} R_{a,d}}{k_{a,w} R_{a,w}} = \frac{1}{(1 - 0.0968SH)(1 + 0.61SH)} \equiv \mu_{\Delta h} \tag{8.57}$$

Therefore

$$\frac{\sqrt{\left(\frac{\Delta h_9^*}{\theta}\right)_{a,d}} - \sqrt{\left(\frac{\Delta h_2}{\theta}\right)_{a,d}}}{\sqrt{\left(\frac{\Delta h_9^*}{\theta}\right)_{a,w}} - \sqrt{\left(\frac{\Delta h_2}{\theta}\right)_{a,w}}} = \frac{\sqrt{\mu_{\Delta h}\left(\frac{\Delta h_9^*}{\theta}\right)_{a,w}} - \sqrt{\mu_{\Delta h}\left(\frac{\Delta h_2}{\theta}\right)_{a,w}}}{\sqrt{\left(\frac{\Delta h_9^*}{\theta}\right)_{a,w}} - \sqrt{\left(\frac{\Delta h_2}{\theta}\right)_{a,w}}} = \sqrt{\mu_{\Delta h}} \tag{8.58}$$

This gives us the thrust humidity correction as follows:

$$
\begin{aligned}
\frac{\left(\frac{F_n}{\delta}\right)_{a,d}}{\left(\frac{F_n}{\delta}\right)_{a,w}} &= \mu_w \sqrt{\mu_{\Delta h}} \\
&= \sqrt{\frac{1 + 0.61SH}{1 - 0.0968SH}} \sqrt{\frac{1}{(1 - 0.0968SH)(1 + 0.61SH)}} \\
&= \frac{1}{1 - 0.0968SH} \equiv \mu_{F_N}
\end{aligned}
\tag{8.59}
$$

$$
\left(\frac{F_n}{\delta}\right)_{a,d} = \frac{1}{1 - 0.0968SH} \left(\frac{F_n}{\delta}\right)_{a,w} \tag{8.60}
$$

8.10 Pressure Humidity Correction

From our discussion of thrust, we used the relationship:

$$
\frac{w_9}{g} v_{9expanded} = A_9(P_9 - P_{s9}) \tag{8.61}
$$

This provides a more general relationship as

$$
\begin{aligned}
&\frac{w}{g} v = A(P_T - P_S) \Rightarrow \\
&\frac{1}{g} \frac{w\sqrt{\theta}}{\delta} \sqrt{2gJ\frac{\Delta h}{\theta}} = A\left(\frac{P_T}{\delta} - \frac{P_S}{\delta}\right)
\end{aligned}
\tag{8.62}
$$

If we let μ_P denote the humidity correction for pressure, then we can assert the following without loss of generality:

$$
\frac{\left(\frac{P_T}{\delta}\right)_{a,d}}{\left(\frac{P_T}{\delta}\right)_{a,w}} = \mu_P = \frac{\left(\frac{P_S}{\delta}\right)_{a,d}}{\left(\frac{P_S}{\delta}\right)_{a,w}} \tag{8.63}
$$

Therefore, if we take the *left hand side* of Eq. (8.63)

$$
\frac{A\left(\frac{P_T}{\delta} - \frac{P_S}{\delta}\right)_{a,d}}{A\left(\frac{P_T}{\delta} - \frac{P_S}{\delta}\right)_{a,w}} = \frac{\left(\frac{P_T}{\delta}\right)_{a,d} - \left(\frac{P_S}{\delta}\right)_{a,d}}{\left(\frac{P_T}{\delta}\right)_{a,w} - \left(\frac{P_S}{\delta}\right)_{a,w}} = \frac{\mu_P\left(\frac{P_T}{\delta}\right)_{a,w} - \mu_P\left(\frac{P_S}{\delta}\right)_{a,w}}{\left(\frac{P_T}{\delta}\right)_{a,w} - \left(\frac{P_S}{\delta}\right)_{a,w}} = \mu_P \tag{8.64}
$$

If we take the right hand side of Eq. (8.61) and expand in the same manner, we find from our derivation for thrust that it is equal to μ_{Fn}. Therefore, $\mu_P = \mu_{Fn}$, i.e.,

$$
\left(\frac{P_T}{\delta}\right)_{a,d} = \left(\frac{1}{1 - 0.0968SH}\right) \left(\frac{P_T}{\delta}\right)_{a,w} \tag{8.65}
$$

8.11 Temperature Humidity Correction

We know, in general, that

$$\frac{P}{P_2} = \left(\frac{T}{T_2}\right)^{\frac{\gamma\eta}{\gamma-1}} \tag{8.66}$$

This implies

$$\mu_P \equiv \frac{\left(\frac{P}{\delta}\right)_{a,d}}{\left(\frac{P}{\delta}\right)_{a,w}} = \frac{\left(\frac{P}{P_2}\right)_{a,d}}{\left(\frac{P}{P_2}\right)_{a,w}} = \frac{\left(\frac{T}{T_2}\right)_{a,d}^{\frac{\gamma_{a,d}\eta}{\gamma_{a,d}-1}}}{\left(\frac{T}{T_2}\right)_{a,w}^{\frac{\gamma_{a,w}\eta}{\gamma_{a,w}-1}}} = \frac{\left(\frac{T}{\theta}\right)_{a,d}^{\frac{\gamma_{a,d}\eta}{\gamma_{a,d}-1}}}{\left(\frac{T}{\theta}\right)_{a,w}^{\frac{\gamma_{a,w}\eta}{\gamma_{a,w}-1}}} \tag{8.67}$$

Taking logs and engaging in some manipulation we obtain

$$\begin{aligned}
\ln(\mu_P) &= \ln\left(\frac{P}{\delta}\right)_{a,d} - \ln\left(\frac{P}{\delta}\right)_{a,w} = \left(\frac{\gamma_{a,d}\eta}{\gamma_{a,d}-1}\right)\ln\left(\frac{T}{\theta}\right)_{a,d} - \left(\frac{\gamma_{a,w}\eta}{\gamma_{a,w}-1}\right)\ln\left(\frac{T}{\theta}\right)_{a,w} \\
&= \left(\frac{\gamma_{a,d}\eta}{\gamma_{a,d}-1}\right)\left\{\underbrace{\ln\left(\frac{T}{\theta}\right)_{a,d} - \ln\left(\frac{T}{\theta}\right)_{a,w}}_{\ln(\mu_T)} + \ln\left(\frac{T}{\theta}\right)_{a,w}\right\} - \left(\frac{\gamma_{a,w}\eta}{\gamma_{a,w}-1}\right)\ln\left(\frac{T}{\theta}\right)_{a,w} \\
&= \left(\frac{\gamma_{a,d}\eta}{\gamma_{a,d}-1}\right)\left\{\ln(\mu_T) + \ln\left(\frac{T}{\theta}\right)_{a,w}\right\} - \left(\frac{\gamma_{a,w}\eta}{\gamma_{a,w}-1}\right)\ln\left(\frac{T}{\theta}\right)_{a,w} \\
&= \left(\frac{\gamma_{a,d}\eta}{\gamma_{a,d}-1}\right)\ln(\mu_T) + \left(\frac{\gamma_{a,d}\eta}{\gamma_{a,d}-1}\right)\ln\left(\frac{T}{\theta}\right)_{a,w} - \left(\frac{\gamma_{a,w}\eta}{\gamma_{a,w}-1}\right)\ln\left(\frac{T}{\theta}\right)_{a,w} \\
&= \left(\frac{\gamma_{a,d}\eta}{\gamma_{a,d}-1}\right)\ln(\mu_T) + \left\{\underbrace{\left(\frac{\gamma_{a,d}\eta}{\gamma_{a,d}-1}\right) - \left(\frac{\gamma_{a,w}\eta}{\gamma_{a,w}-1}\right)}_{\approx 0}\right\}\ln\left(\frac{T}{\theta}\right)_{a,w} \\
&\approx \left(\frac{\gamma_{a,d}\eta}{\gamma_{a,d}-1}\right)\ln(\mu_T) \\
&= \ln(\mu_T)^{\left(\frac{\gamma_{a,d}\eta}{\gamma_{a,d}-1}\right)} \approx \ln(\mu_T)^{3.5} \Rightarrow \\
\mu_P &\approx \mu_T^{3.5}
\end{aligned} \tag{8.68}$$

Therefore

$$\mu_T \equiv \frac{\left(\frac{T}{\theta}\right)_{a,d}}{\left(\frac{T}{\theta}\right)_{a,w}} \approx \left(\frac{1}{(1-0.0968SH)}\right)^{1/3.5} \tag{8.69}$$

Table 8.2 Summary of humidity corrections

Parameter	Symbol	Humidity correction	Relationship
Spool speed	$\frac{N}{\sqrt{\theta}}$	$\frac{1}{\sqrt{[1-0.0968SH][1+.61SH]}}$	μ_N
Air flow	$\frac{w_a\sqrt{\theta}}{A\delta}$	$\sqrt{\frac{1+0.61SH}{1-.0968SH}}$	μ_{w_a}
Enthalpy	$\frac{\Delta h}{T}$	$\frac{1}{(1-0.0968SH)(1+0.61SH)}$	$\mu_{\Delta h}$
Δ Temperature	$\frac{\Delta T}{T}$	$\frac{(1+0.852SH)}{(1-0.0968SH)(1+0.61SH)}$	$\frac{cp_{a,w}}{cp_{a,d}}\mu_{\Delta h}$
Horsepower	$\frac{SHP}{\sqrt{\theta\delta}}$	$\frac{1}{\sqrt{(1+0.61SH)(1-0.0968SH)^{1.5}}}$	$\frac{cp_{a,d}}{cp_{a,w}}\mu_w\mu_{\Delta T/T}$
Torque	$\frac{Q}{\delta}$	$\frac{1}{(1-0.0968SH)}$	$\frac{\mu_{SHP}}{\mu_N}$
Acceleration	$\frac{\dot{N}}{\delta}$	$\frac{1}{(1-0.0968\,SH)}$	μ_Q
Fuel Flow	$\frac{W_f}{\sqrt{\theta\delta}}$	$\frac{1}{(1-0.0968SH)^{1.5}\sqrt{(1+0.61SH)}}$	$\mu_{w_f}=\mu_w\mu_{\Delta h}$
Thrust	$\frac{F_n}{\delta}$	$\frac{1}{(1-0.0968SH)}$	μ_{F_N}
Pressure	$\frac{P}{\delta}$	$\frac{1}{(1-0.0968SH)}$	$\mu_P=\mu_{F_N}$
Temperature	$\frac{T}{\theta}$	$\left(\frac{1}{(1-0.0968SH)}\right)^{1/3.5}$	μ_T

yielding

$$\left(\frac{T}{\theta}\right)_{a,d}=\left(\frac{T}{\theta}\right)_{a,w}\left(\frac{1}{(1-0.0968SH)}\right)^{1/3.5} \tag{8.70}$$

A summary of the humidity effects derived in this Chapter appears in the Table 8.2 above.

8.12 Numerical Quantification of Humidity Effects

The corrections for humidity derived in this Chapter (and summarized in Table 8.3) are multiplicative factors that can be applied to various gas path parameters to determine the equivalent value that would be seen under dry air conditions (Fig. 8.1). The tables and plots below indicate the relative magnitude effect for the common parameters as multiplicative factors as a function of specific humidity.

The table and plot below depict the same effects given as percent change experienced for each gas path parameter (Table 8.4 and Fig. 8.2).

Table 8.3 Humidity correction multiplicative factors

Humidity corrections										
SH	Spool speed	Air flow	Temperature ΔT/T	Shaft horsepower	Torque	Spool accelerat	Fuel flow	Thrust	Pressure	Temp
0.00000	1	1	1	1	1	1	1	1	1	1
0.00500	0.99872	1.00177	1.00169	0.99920	1.00048	1.00048	0.999204	1.000484	1.000484	1.000138
0.01000	0.99745	1.00353	1.00338	0.99841	1.00097	1.00097	0.998413	1.000969	1.000969	1.000277
0.01500	0.99618	1.00529	1.00506	0.99763	1.00145	1.00145	0.997628	1.001454	1.001454	1.000415
0.02000	0.99492	1.00706	1.00673	0.99685	1.00194	1.00194	0.996849	1.00194	1.00194	1.000554
0.02500	0.99366	1.00882	1.00840	0.99607	1.00243	1.00243	0.996075	1.002426	1.002426	1.000693
0.03000	0.99242	1.01058	1.01006	0.99531	1.00291	1.00291	0.995306	1.002912	1.002912	1.000831
0.03500	0.99117	1.01233	1.01172	0.99454	1.00340	1.00340	0.994543	1.0034	1.0034	1.00097
0.04000	0.98994	1.01409	1.01337	0.99379	1.00389	1.00389	0.993785	1.003887	1.003887	1.001109

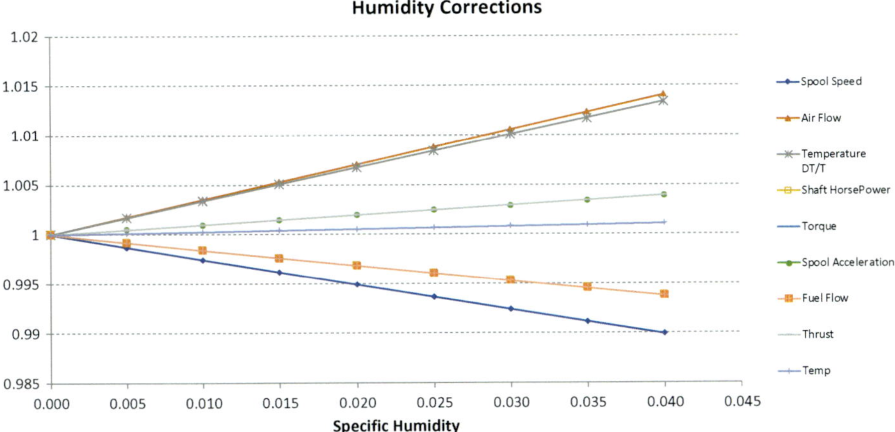

Fig. 8.1 Humidity correction multiplicative factors

Table 8.4 % Humidity correction factors

Humidity corrections (%)										
SH	Spool speed	Air flow	Temperature Δt/t	Shaft horsepower	Torque	Spool acceleration	Fuel flow	Thrust	Pressure	Temp
0.00000	0.00	0.00	0.00	0.00	0.00	0.00	0.00	0.00	0.00	0.00
0.00500	−0.13	0.18	0.17	−0.08	0.05	0.05	−0.08	0.05	0.05	0.01
0.01000	−0.26	0.35	0.34	−0.16	0.10	0.10	−0.16	0.10	0.10	0.03
0.01500	−0.38	0.53	0.51	−0.24	0.15	0.15	−0.24	0.15	0.15	0.04
0.02000	−0.51	0.71	0.67	−0.32	0.19	0.19	−0.32	0.19	0.19	0.06
0.02500	−0.63	0.88	0.84	−0.39	0.24	0.24	−0.39	0.24	0.24	0.07
0.03000	−0.76	1.06	1.01	−0.47	0.29	0.29	−0.47	0.29	0.29	0.08
0.03500	−0.88	1.23	1.17	−0.55	0.34	0.34	−0.55	0.34	0.34	0.10
0.04000	−1.01	1.41	1.34	−0.62	0.39	0.39	−0.62	0.39	0.39	0.11

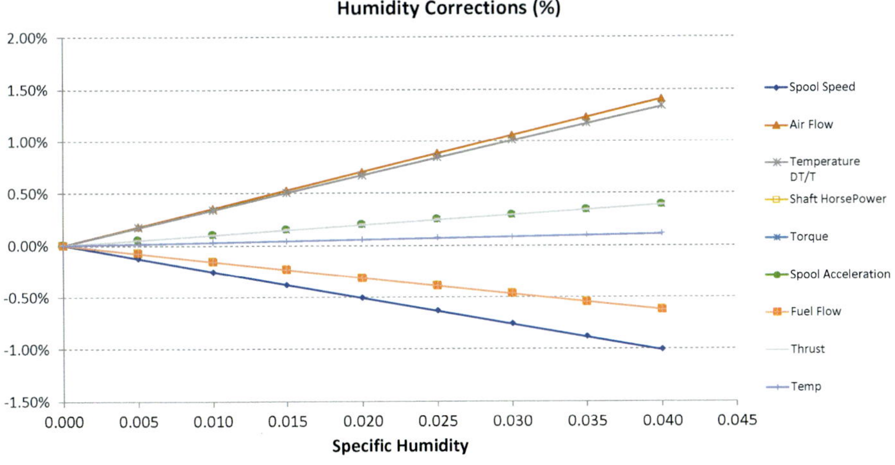

Fig. 8.2 Humidity correction % effect

Appendix: Summary of Corrections

In this appendix we will summarize the corrections derived in the previous chapters. The format we will use is to list corrections by gas path parameter. This will include for each parameter, its Standard Day Correction along with any refinements, time derivative correction and humidity correction as applicable.

Rotational Speed

Standard

$$N_c = \frac{N}{\sqrt{\theta}}$$

Refinement

$$N_c = \frac{N}{\theta^{0.48\,\alpha_T}} = \frac{N}{\theta^{0.48}}$$

Derivative

$$\dot{N}_c = \frac{\dot{N}}{\delta}$$

$$\dot{N}_c = \frac{\dot{N}}{\delta\,\theta^{1.12\alpha_T+\alpha_{w_a}-0.48}}$$

$$\ddot{N}_c = \frac{\ddot{N}\sqrt{\theta}}{\delta^2}$$

A. J. Volponi, *Gas Turbine Parameter Corrections*,
https://doi.org/10.1007/978-3-030-41076-6

Humidity

$$\left(\frac{N}{\sqrt{\theta}}\right)_{a,d} = \left(\frac{N}{\sqrt{\theta}}\right)_{a,w} \frac{1}{\sqrt{[1 - 0.0968SH]\,[1 + 0.61SH]}} = \left(\frac{N}{\sqrt{\theta}}\right)_{a,w} \mu_N$$

$$\left(\frac{\dot{N}}{\delta}\right)_{a,d} = \left(\frac{\dot{N}}{\delta}\right)_{a,w} \left[\frac{1}{(1 - 0.0968SH)}\right] = \left(\frac{\dot{N}}{\delta}\right)_{a,w} \mu_{\dot{N}}$$

Temperature

Standard

$$T_c = \frac{T}{\theta}$$

Refinement

$$T_{sc} \approx \frac{T_s}{\theta^{1.0033\,\alpha_T}}$$

Derivative

$$\dot{T}_c = \frac{\dot{T}}{\delta\sqrt{\theta}}$$

Humidity

$$\left(\frac{\Delta T}{T_s}\right)_{a,d} = \frac{(1 + 0.852SH)}{(1 + 0.61SH)(1 - 0.0968SH)} \left(\frac{\Delta T}{T_s}\right)_{a,w}$$

$$\left(\frac{\Delta h}{T}\right)_{a,d} = \frac{1}{(1 - 0.0968SH)(1 + 0.61SH)} \left(\frac{\Delta h}{T}\right)_{a,w}$$

$$\left(\frac{\Delta T}{T}\right)_{a,d} = \frac{(1 + 0.852SH)}{(1 - 0.0968SH)(1 + 0.61SH)} \left(\frac{\Delta T}{T}\right)_{a,w}$$

Pressure

Standard

$$P_c = \frac{P}{\delta}$$

Refinement

$$P_c \approx \frac{P}{\delta\theta^{\frac{\gamma\eta}{\gamma-1}(\alpha_T-1)+.041}\left[\left(\frac{\gamma\eta}{(\gamma-1)^2}\right)\ln\left(\frac{T^*}{T}\right)\right]}$$

Derivative

$$\dot{P}_c = \frac{\dot{P}\sqrt{\theta}}{\delta^2}$$

Humidity

$$\left(\frac{P_T}{\delta}\right)_{a,d} = \left(\frac{1}{1-0.0968SH}\right)\left(\frac{P_T}{\delta}\right)_{a,w}$$

Air Flow

Standard

$$w_{ac} = \frac{w_a\sqrt{\theta}}{\delta}$$

Refinement

$$w_{ac} = \frac{w_a}{A\delta\theta^{-1/2\,(1+0.041\,x\,\alpha_T)}} = \frac{w_a}{A\delta\theta^{\alpha_{wa}}}$$

Derivative

$$\dot{w}_{ac} = \frac{\dot{w}_a\theta}{\delta^2}$$

Humidity

$$\left(\frac{w\sqrt{\theta}}{A\delta}\right)_{a,d} = \left(\frac{w\sqrt{\theta}}{A\delta}\right)_{a,w}\left(\sqrt{\frac{1+0.61SH}{1-.0968SH}}\right) = \left(\frac{w\sqrt{\theta}}{A\delta}\right)_{a,w}\mu_w$$

Fuel Flow

Standard

$$w_{fc} = \frac{w_f}{\delta\sqrt{\theta}}$$

Refinement

$$w_{fc} = \frac{w_f}{\delta\theta^{1.12\alpha_T(1+FAR)+\alpha_{wa}}} = \frac{w_f}{\delta\theta^{\alpha_{w_f}}}$$

Derivative

$$\dot{w}_{fc} = \frac{w_f}{\delta^2}$$

Humidity

$$\left(\frac{w_f}{\sqrt{\theta}\delta}\right)_{a,d} = \left(\frac{w_f}{\sqrt{\theta}\delta}\right)_{a,w} \left\{\frac{1}{(1-0.0968SH)^{1.5}\sqrt{(1+0.61SH)}}\right\}$$
$$= \left(\frac{w_f}{\sqrt{\theta}\delta}\right)_{a,w} \mu_{w_f}$$

Horsepower

Standard

$$HP_c = \frac{HP}{\delta\sqrt{\theta}}$$

Refinement

$$HP_c = \frac{HP}{\delta\theta^{(1.12\alpha_T+\alpha_{wa})}}$$

Derivative

$$\dot{H}P_c = \frac{\dot{H}P}{\delta^2}$$

Humidity

$$\left(\frac{HP}{\sqrt{\theta}\delta}\right)_{a,d} = \left(\frac{HP}{\sqrt{\theta}\delta}\right)_{a,w} \frac{1}{\sqrt{(1+0.61SH)(1-0.0968SH)^{1.5}}} = \left(\frac{HP}{\sqrt{\theta}\delta}\right)_{a,w} \mu_{HP}$$

Torque

Standard

$$Q_c = \frac{Q}{\delta}$$

Refinement

$$Q_c = \frac{Q}{\delta\theta^{1.12\alpha_T+\alpha_{w_a}-0.48}}$$

Derivative

$$\dot{Q}_c = \frac{\dot{Q}\sqrt{\theta}}{\delta^2}$$

Humidity

$$\left(\frac{Q}{\delta}\right)_{a,d} = \left(\frac{Q}{\delta}\right)_{a,w}\left[\frac{1}{(1-0.0968SH)}\right] = \left(\frac{Q}{\delta}\right)_{a,w}\mu_Q$$

Metal Temperature

Standard

$$T_{mc} = \frac{T_m\delta^{0.2}}{\theta^{1.24}} \approx \left(\frac{\delta}{\theta}\right)^{0.2}\frac{T_m}{\theta} \equiv \alpha_m\frac{T_m}{\theta}$$

Derivative

$$\dot{T}_{mc} = \frac{\dot{T}_m}{\theta^{0.74}\delta^{0.8}}$$

Thrust

Standard

$$F_{nc} = \frac{F_n}{\delta}$$

Derivative

$$\dot{F}_{nc} = \frac{\dot{F}_n\sqrt{\theta}}{\delta^2}$$

Humidity

$$\left(\frac{F_n}{\delta}\right)_{a,d} = \frac{1}{1 - 0.0968SH}\left(\frac{F_n}{\delta}\right)_{a,w}$$

Time Constant

Standard

$$\tau_c = \left(\frac{\delta}{\sqrt{\theta}}\right)\tau$$

Fuel Air Ratio

Standard

$$FAR_c = \frac{FAR}{\theta}$$

Derivative

$$F\dot{A}R_c = \frac{F\dot{A}R}{\delta\sqrt{\theta}}$$

Thrust Specific Fuel Consumption

Standard

$$TSFC_c = \frac{TSFC}{\sqrt{\theta}}$$

Derivative

$$TS\dot{F}C_c = \frac{TS\dot{F}C}{\delta}$$

General

Standard

$$\begin{aligned} X_c &= \frac{X}{\theta^a \delta^b} \\ (X+Y)_c &= X_c + Y_c \\ (XY)_c &= X_c Y_c \\ \left(\sqrt{X}\right)_c &= \sqrt{X_c} \\ \left(X^\beta\right)_c &= X_c^\beta \\ \dot{X}_c &= \frac{\sqrt{\theta}}{\delta} \frac{d}{dt}(X_c) \end{aligned}$$

References

1. Warner, D. F., & Auyer, E. L. (1945). Contemporary jet-propulsion gas turbines for aircraft. *Mechanical Engineering, 67*(11), 1945.
2. Sanders, N. D. (1946). *Performance parameters for jet-propulsion engines* (NACA Technical Note No. 1106).
3. Kurzke, J. (2003). Model based gas turbine parameter corrections. In *Proceedings of 2003 ASME TURBO EXPO: Power for Land, Sea, & Air, Paper GT2003-38234*, June 16-19, 2003, Atlanta, Georgia, USA.
4. Assadi, M., et al. A novel correction technique for simple gas turbine parameters. In *Proceedings of 2003 ASME TURBO EXPO: Power for Land, Sea, & Air, Paper GT2001-0009*, June 4-7, 2001, New Orleans, USA.
5. Buckingham, E. (1914). On physically similar systems: Illustrations of the use of dimensional equations. *Physical Review, 4*, 345.
6. Volponi, A. J. (1999). Gas turbine parameter corrections. *Transactions of the ASME, Journal of Engineering for Gas Turbines and Power, 121*(4), 613–621.
7. Shepherd, D. G. (1949). *An introduction to the gas turbine* (p. 254). London: Constable and Company, Ltd.
8. Samuel, J. C., & Gale, B. M. (1950, June). *Effect of humidity on performance of turbojet engines*. National Advisory Committee for Aeronautics, Technical Note 2119.
9. Mathioudakis, K., & Tsalavoutas, A. (2003). Uncertainty reduction in gas turban performance diagnostics by accounting for humidity effects. In *Proceedings of 2003 ASME TURBO EXPO: Power for Land, Sea, & Air, Paper GT2001-0010*, June 4-7, 2001, New Orleans, USA.
10. AGARD. (1995, September). *Recommended practices for the assessment of humidity effects of atmospheric water ingestion on the performance and operability of gas turbine engines* (AGARD Advisory Report No. 332).
11. Samuels, J. C., & Gale, B. M. (1950). *Effect of humidity on performance of turbojet engines, NACA Technical Note 2119*. Cleveland, OH: Lewis Flight Propulsion Laboratory.
12. Hanachi, H., et al. (2015). Effects of the intake air humidity on the gas turbine performance monitoring. In *Proceedings of 2015 ASME TURBO EXPO: Power for Land, Sea, & Air, Paper GT2015-43026*, June 15-19, 2015, Montreal, Canada.
13. Hill, P., & Petersen, C. (1992). *Mechanics and thermodynamics of propulsion*. Prentice Hall: Pearson Education, Inc.
14. Oates, G. (1997). *Aerothermodynamics of gas turbine and rocket propulsion*. AIAA Education Series.
15. Bridgman, P. W. (1922). *Dimensional analysis*. New Haven: Yale University Press. ISBN 978-0-548-91029-0.

A. J. Volponi, *Gas Turbine Parameter Corrections*,
https://doi.org/10.1007/978-3-030-41076-6

Index

A. J. Volponi, *Gas Turbine Parameter Corrections*,
https://doi.org/10.1007/978-3-030-41076-6

GPSR Compliance

The European Union's (EU) General Product Safety Regulation (GPSR) is a set of rules that requires consumer products to be safe and our obligations to ensure this.

If you have any concerns about our products, you can contact us on ProductSafety@springernature.com

In case Publisher is established outside the EU, the EU authorized representative is:

Springer Nature Customer Service Center GmbH
Europaplatz 3
69115 Heidelberg, Germany

Batch number: 10372213

Printed by Printforce, the Netherlands